CONTAINER ATLAS

A PRACTICAL GUIDE TO CONTAINER ARCHITECTURE

SLAWIK, BERGMANN, BUCHMEIER, TINNEY (Eds.)

gestalten

FOREWORD

Starting out from the original intention of collecting my own buildings and projects under the title of "*Container*Architecture", the dramatic increase in the number of container buildings worldwide led to the development of the idea of producing an atlas that would serve as a manual and present an overview of the state of "*Container*Architecture". With the uninterrupted boom in container projects, interest in this subject and the demand for information regarding aspects of the design and construction of containers have also increased.

As the head of the "Experimental Design and Construction" department within the Faculty of Architecture and Landscape Sciences at Leibniz University Hanover, I was able to draw on my academic and research staff as an expert team to help with the creation of the first container atlas in the world. For a number of years now, container architecture has been the explicit focus of our research activity. This has been accompanied by student designs and seminars with the deliberate aim of increasing our store of "container knowledge".

I would like to sincerely thank the research workers Julia Bergmann, Matthias Buchmeier and Sonja Tinney for their enthusiasm and unwavering dedication. Our tutors Anja Iffert and Lisa Lüdke also deserve thanks for their contribution.

We would also like to thank the Dutch engineer Douwe de Jong for his guest article that provides more detailed information on the structural aspects of building using freight containers. Special thanks go to the entrepreneur Dr. Christian Seidel, the architectural sociologist Heino Sandfort and the architect Carsten Wiewiorra for their valuable suggestions regarding the manuscript. Last but not least, we would like to expressly thank the architects themselves and their photographers for their willingness to allow us to use their material.

We invite you, the reader, to take a trip through the world of alternative, "non-bourgeois" architecture. We hope that the *Container Atlas* will help to increase awareness and understanding of this subject, and that this work will serve as a manual for building using containers for specialists and interested laypeople alike.

Prof. Han Slawik

CONTENTS

01	THE HISTORY OF THE SHIPPING CONTAINER	6
	Background	
02	FROM THE CONTAINER TO AN ARCHITECTURE	8
	Introduction	
03	USE OF CONTAINERS	11
04	CONTAINERS AS BUILDING MODULES	14
	Fundamentals	
	Definition	14
	Principles	15
	Foundation	16
	Construction physics	17
	Construction law	17
	Economic aspects	18
	Ecological aspects	18
	International aspects	18
05	CONTAINERS AS BUILDING BLOCKS	20
	Type 1 / Freight containers	21
	Definition	22
	Standard	22
	Dimensions	22
	Weights and loads	23
	Types	24
	Costs	25
	Transportation	26
	Production	26
	Construction	27
	Technical equipment	28
	Security	29
	Type 2 / Building containers	30
	Definition	30
	Standard	30
	Use	30
	Types	32
	Dimensions	32
	Construction	33
	Production	34
	Transportation	34
	Variants	36
	Construction physics	36
	Economic aspects	36
	Special forms	37
	Type 3 / Container frames	38
	Type 3A: Modular frame systems	38
	Definition	38
	Development	38
	Use	38
	Dimensions	39
	Production	39
	Construction	40
	Transportation	40
	Construction physics	40
	Economic aspects	41
	Ecological aspects	41
	Type 3B: Container frame systems	42
	Definition	42
	Dimensions	42
	Use	42
	Construction	42
	Technical fittings	44
	Economic aspects	44
	Ecological aspects	44
	Remarks	45
06	STRUCTURAL DESIGN ASPECTS: FREIGHT CONTAINERS	46
	Guest contribution from the engineer Douwe de Jong	
07	PROJECTS	48
	List of projects	
	Projects	50
08	OUTLOOK	238
09	APPENDIX	239
	Timeline	240
	Index of Images	242
	Bibliography	243
	Containers in general	243
	Freight containers	243
	Building containers	243
	Container frames	243
	List of companies	244
	Freight containers	244
	Building containers	244
	Container frames	244
	Organizations	245
	Regulatory works	245
	Standards	245
	Guidelines	245
	Glossary	246
	Index of projects	248
	Editors	255
	Imprint	256

01 BACKGROUND
THE HISTORY OF THE SHIPPING CONTAINER

While young Malcom McLean sat behind the wheel of his truck at the pier in Hoboken, New Jersey, waiting hours for bales of cotton to be unloaded from his truck, he couldn't possibly have known what an impact the idea he had dreamt about that night would have on the world. Rather, he was bored and frustrated because as the owner of a trucking company, every wasted hour cost him hard cash.

Malcom Purcell McLean was born in North Carolina in 1913 and grew up on a farm during the Great Depression. The hard physical work on the farm made him realize from early on, that he wanted to do something different with his life. With only a high school diploma, he left the farm and started to work at a gas station in a town close by. When a construction manager came in one day looking for a driver with a truck, McLean simply went to the local Ford dealer's, bought a used pick-up for $120 and started working for him. He soon realized that he would be able to make good money with a haulage company and started to set up a fleet of trucks. Back then it was hard to imagine that "McLean Trucking Co." would become the second largest trucking company in the USA with 1770 vehicles and 32 terminals. But on his way to the top, there were a couple of setbacks such as the heavy ice storms in 1936 that caused accidents and the cancellation of orders, thus forcing the successful start-up entrepreneur to get back behind the wheel himself. On the aforementioned tour in the following year, he had a revolutionary idea while he was impatiently waiting for the longshoremen to unload his freight of cotton from the truck onto the ship, bale by bale, with a hook. How easy would it be if one could only take off the whole truck mounting and then, at the ship-to location, just put it on another truck or freight train?

The idea of transporting goods in containers in order to facilitate the process of loading and unloading wasn't new at that time. Since the beginning of the past century, metal boxes had been used for transportation of goods. On the route between Dover and Calais, vehicle-comprehensive modules were even already in use, but none of these systems had made it out of their niche. McLean was seeking a universal solution. The idea of introducing standardized containers the size of a truck's loading space suitable for all major means of transportation was born at that moment, but the path to implementing this system was very long and difficult. None of the sectors whose participation would have been required for the inter-modal container's success were willing to venture out on this new trail. The traditional shipping industry, in particular, was reluctant to accept the bare sheet metal box. Apart from the fact that they did not believe in the concept of mechanically unloading goods, a system such as the one proposed would completely turn existing logistics upside-down. The handling of individually packaged goods was very lucrative for shipping companies because the transport price was composed of many different parameters such as quantity, size, weight and value as well as fees for special handling. Basically, contractors did not have any price transparency or alternatives for sending their goods. And here was this Malcom McLean telling them that with his new system, ships wouldn't have to stay in harbors any longer than a couple of hours because machines were going to do the unloading. Instead of 20 longshoremen, only one would be needed. Transportation prices would become fairer and more transparent; freight would be handled in bundles and in closed boxes so that the chances of losing goods or having them stolen would be minimized. This whole new way of handling goods would lead to a loss of jobs as well as the cherished harbor romanticism that had a great appeal to many workers in the shipping industry. It would mean the end of extensive shore leaves and the vivid quayside bar culture; no more chances to secretly pocket a carton of cigarettes, a tin of coffee or a few oranges. It was unimaginable that a country boy wanted to eliminate all that.

Like any revolutionary idea that is born, its realization couldn't be stopped. At best, the process could be slowed down—in the case of the shipping container, it took about 20 years. First of all, McLean had to rescue his trucking company. He took his sister Clara and his brother Jim on board and started building terminals throughout the country. Every employee he hired had to begin by going on the road for half a year, because McLean was convinced that only those who had driven themselves and thus had learned how to handle freight, change oil, and maintain the engine were really able to evaluate freight prices. He kept developing new training programs because he regarded well-trained employees as the key to a successful business. Part of his philosophy was the rule that no one was allowed to give trucks names or place nametags inside the driver's cab. This was because he didn't want his drivers to become attached to a particular vehicle as they would then give "their" truck special treatment and wouldn't want to drive other trucks of the fleet. He believed that a company that allowed its employees to develop a personal relationship to one particular truck couldn't run efficiently. Thus, instead of having names, every vehicle at McLean Trucking was starkly numbered. Whenever Malcom McLean launched into something, sensitivities had to yield to profitability.

In 1965, McLean had enough money to make his dream of a universal freight container for ships, trucks and trains come true. Since he still couldn't expect any support from the various transport branches, he decided to become a shipping company owner himself. He bought the Pan-Atlantic Steamship Corporation after fighting stubbornly, and eventually successfully, for a loan that actually exceeded the bank's limit. His competitors were incensed: they sued him, citing the then valid anti-monopoly law, which prohibited companies from operating sea and land transport at the same time. McLean had to choose between his new shipping company and his very successful trucking business, which he had established in tedious work over so many years. Without hesitation, he gave up trucking because he was so obsessed with his new business plan and convinced of its success. When the first container ship, the "Ideal X", with on-board loading bridges, left the harbor of Newark, New Jersey that same year, experts were certain that this was nothing but some very expensive pipe dream that was not going to survive the reality check. His own crew had difficulties with the innovations in their traditional métier, too. When their boss suggested abolishing the ships' names and giving them numbers just as he had done with the trucks, there was almost mutiny. But it was this very novel approach that made him the revolutionary he was. What distinguished him from other ship owners, who mainly thought about shipping, or the truck business owners, who only thought about trucking, was that he was concerned about nothing but the freight. When he opted for the shipping branch and against trucking, he didn't switch to a different industry; he only changed the means of transportation for his freight.

In 1960, the Pan-Atlantic Steamship Corporation was renamed Sealand Services. The new, simple and more direct name mirrored McLean's philosophy that had no room for maritime nostalgia; this company was all about efficiency and success. McLean didn't bother to build new ships—similarly to his former trucking business, he used cargo ships and converted them to hold containers because it saved time and money. The beginnings of Sealand Services were rough: the enterprise had to circumnavigate bankruptcy several times, but all the young employees that McLean had wisely selected with the help of his siblings were highly motivated and worked hard and enthusiastically for the new idea. Within the following years, more and more container ships landed in US American harbors. While followers commuted between the West Coast and Hawaii, Sealand Services covered the routes along the East Coast. Nevertheless, the old established shipping companies still did not believe in his long-term success, especially in international overseas traffic. When McLean announced that soon he was going to send extremely fast container ships across the Atlantic at intervals of only two days, nobody took him seriously.

Many people are still not quite aware of the great extent to which the freight container influences our life today, even though it has changed consumer behavior dramatically in most parts of the world. Sand shrimp from the North Sea is only available at discount prices in German supermarkets because McLean's containers ship them from Bremen to Mexico and back at a very low cost so that Mexican workers who have probably never tasted sand shrimp in their lives can shell them for a fraction of the wage that a German worker would cost. Today, hardly any final product is assembled in one place anymore; every single component of a thermos flask is produced and added at a different manufacturing plant somewhere in the world. Nations in the most remote corners of the globe now have the opportunity to participate in the world market because no route of transport is as cheap as the one by sea. The introduction of the shipping container has turned oceans into maritime highways.

Back then, the world wasn't waiting for McLean's invention He had to make himself and his ideas known, and the clever businessman sniffed out his great chance during the Vietnam War. Shortly after its outbreak, the US Army experienced difficulty in getting supplies to its soldiers; freight ships loaded with food and military equipment were jammed up outside the harbor of Hanoi because it took so long to unload. With his container system, McLean had the perfect solution for the problem but when he wanted to present it, he wasn't even given an appointment at the Pentagon. Without further ado, he traveled to Arlington, Virginia himself and waited in front of the supply officer's door in order to intercept him on the way to work at seven o'clock in the morning. One last time, his revolutionary idea was greeted with nothing but skepticism. How could this man claim to unload his ships within only 24 hours when everyone else needed at least several days? A free tour arranged right on the spot finally convinced the US government, and a few years later, the rest of the world followed suit. Since the government paid for both ways, back and forth, and McLean didn't want to waste a single traveled sea mile, he soon started organizing the transport of goods from Japan, Hong Kong and Taiwan for the return trip, thereby causing not only an invasion of Asian foods, toys and electronic devices to the US, but also laying the foundation for globalization.

Finally, the farm boy from North Carolina had won the long battle against the shipping industry and at the same time, eliminated almost all manpower from the process of bringing goods to the people. The few dockworkers left in the harbors have no idea what commodities they are loading and unloading every day at their computer-operated terminals, with the help of driverless vehicles. There is almost nothing that cannot be shipped in a standardized metal box.

02 INTRODUCTION
FROM THE CONTAINER TO AN ARCHITECTURE

The freight container as a storage and transport vessel for goods revolutionized the transport sector in the last century. Instead of loading and unloading using port cranes or ship cranes, modern global transport is now based on container ships and container terminals. Freight containers can be found all over the world—from the Antarctic to the tropical rain forest.

The "container revolution" began in the 20th century. Malcolm McLean, a former trucking entrepreneur, was one of the first to implement the idea of standardized containers and developed these in 1956 for his trucking empire in the USA in the form of 35-foot-long boxes that could be loaded onto ships. At around the same time, the military was also instrumental in promoting the spread of containers. Back as early as the Second World War, the US Army had already employed rectangular containers as a solution to major logistical problems in crisis areas. The US Army also developed 20-foot boxes that could be transported by water and by land. These developments were then adopted worldwide in the 1960s. Standardization of freight containers in the 1970s, in line with the ISO (International Standards Organization) standard, helped put in place the prerequisites for the worldwide dominance of containers.

The main technical details regarding containers were specified in this ISO standard. The maximum dimensions of containers are mainly determined by the transport conditions, as the locally applicable road traffic regulations prescribe the maximum size of container trailer chassis. 20-foot (6 meters nominal size) and 40-foot (12 meters nominal size) freight containers have become established today from among the various different lengths of container types available. They have a standard width of 8 feet (≈2.4 meters) and various heights: standard cube with 8.5 feet (≈2.6 meters), low cube (rare) with 8 feet (≈2.4 meters) and (increasingly) high cube with 9.5 feet (≈2.9 meters). Transportation and lifting equipment the world over, is tailored to match these dimensions. One 40-foot container or two 20-foot containers can be transported on a single chassis. Alongside the standard containers with lengths of 20 feet and 40 feet, there are also variants and special designs for various purposes: ventilated, cooling, open-top, open-side, bulk, tank, and platform/flat containers.

A standard container consists of a steel construction with standardized special profiles and load-bearing walls. Today these steel containers are generally made of slow-rusting COR-TEN steel. However, there are also other variants—for example: containers with non-load-bearing wooden wall fillings (plywood containers), those made of aluminum (half the weight, double the price), or more recently, containers made of plastics that have a supporting steel frame.

The load-bearing capacity of containers is also specified in the standard. The containers must be able to withstand deformations in line with specified standard values, and must be fully sealed. Because they are sealed, containers do not sink initially at sea: shipping accidents have resulted in around 30,000 containers currently floating aimlessly on the world's seas—they often lie just under the surface of the water, a potential nightmare for anyone hoping to sail around the world.

From the container to an architecture

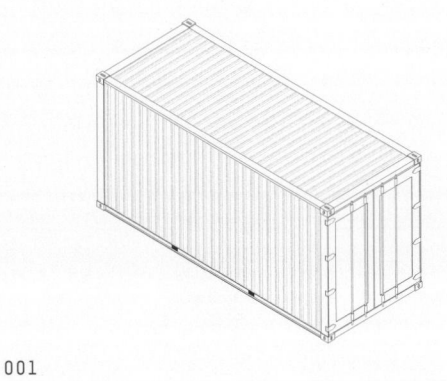

001

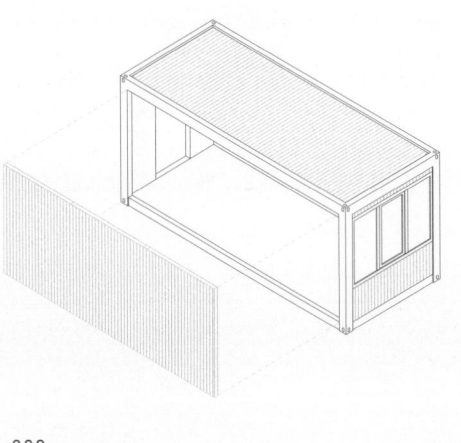

002

Freight containers are extremely stable: a 20-foot container weighs 2.4 metric tons, can take a load of 24 tons, and can be stacked eightfold. A 40-foot container weighs 4.5 metric tons, but can only take a load of 30 tons as the load is still only supported by four corners, thus limiting the load-bearing capacity.

Freight containers are also very inexpensive: a new 20-foot container costs around €2,500 and a used one around €1,300. Most containers are produced in Asia and have generally been used once for freight already.

The supporting structure and weather-protective shell is thus available for a price of approximately €200 per square meter or around €60 per cubed meter, a price that is unbeatable when compared to the costs of erecting a conventional building.

Freight containers were first put to new uses in unconverted form. For example, they were used as tool sheds or storage space. The next step was to use freight containers for other purposes, such as for living in, and to convert them accordingly. The load-bearing capacity of shipping containers is very high, but comprehensive conversion work on the basic structure can impair their static properties. As conversion work can be very cost-intensive, the amount of work carried out is often limited by the amount of finance available. Freight containers are almost always used in architecture for temporary construction purposes. Minimum building physics requirements, mainly relating to heat insulation, can be difficult to achieve with unmodified freight containers, however. Containers are favored in event architecture because of their image.

001 | Freight container system

Building containers have become very widespread in the construction industry and are used mainly in Europe. These are containers with a significantly lighter construction that are used as offices or for commercial or housing purposes. They were originally also produced in ISO dimensions, but later on developed their own sizing systems and were fitted with specific transport features. These containers are familiar as construction-site offices, emergency housing for asylum seekers, accommodation in disaster areas, etc. From a statics point of view, it is possible to stack these containers to create up to three stories. They can be stacked up to four stories in exceptional cases, but reinforced constructions must then be used. More stringent building physics requirements, such as those necessary for permanent constructions, can only be fulfilled with increased labor and costs, meaning that building containers are generally only equipped to a minimum standard. They are placed in rows and stacked, and this is generally done without any regard for architectural form. Some manufacturers do make an effort to give their products a certain quality level by providing more generous facade solutions and by carefully selecting and processing their surface materials—with a commendable degree of success, too. Since these systems are not compatible with similar products from other manufacturers, we refer to these systems as *closed systems*.

002 | Building container system

With all of the systems described so far, the building containers must be custom-manufactured if they are to be joined to produce larger combined spaces. Since fittings components always contribute to load-bearing too, the limits of what is statically permissible can be quickly exceeded if wall, ceiling or floor fillings are omitted. When a number of modules are arranged in rows or stacked, these components double up and are redundant, as there are always two walls, ceilings or floors beside or on top of each other. In principle, the building containers can be reused after being disassembled, but excessive costs may be involved in adapting them to meet new requirements. The sustainability of this building system is thus limited, as the components may need to be processed and treated before being reused.

The principles behind building containers have been transferred to module frames, which can be manufactured in any size, independently of the ISO dimensions system. The fillings also perform a load-bearing function in this building system. This provided the impetus for Professor Han Slawik to develop his own building module, the container frame, and to propose a systematic separation between the load-bearing frame and non-load-bearing fillings. The container frame was the first system to strictly adhere to the idea of adding non-load-bearing fillings to the supporting frame, thus separating the supporting structure and the fittings. There are similar systems available from various manufacturers, but they rely on the shell construction making a contribution to structural strength. This container frame building module has been further developed over the course of various studies, competitions and patent applications. The non-load-bearing fillings are interchangeable in this system, thus guaranteeing maximum flexibility and variability for these modules. With the thermally separated frames and fittings on the frame level and outer premounted shells as an alternative, the building physics requirements could be optimized to a significant extent. The implementation of this type of "pure" system certainly involves more labor and is more costly, but also has the significant advantage that all components created according to a standard, universal system of dimensions are interchangeable and can be reused. Ideally, the components should

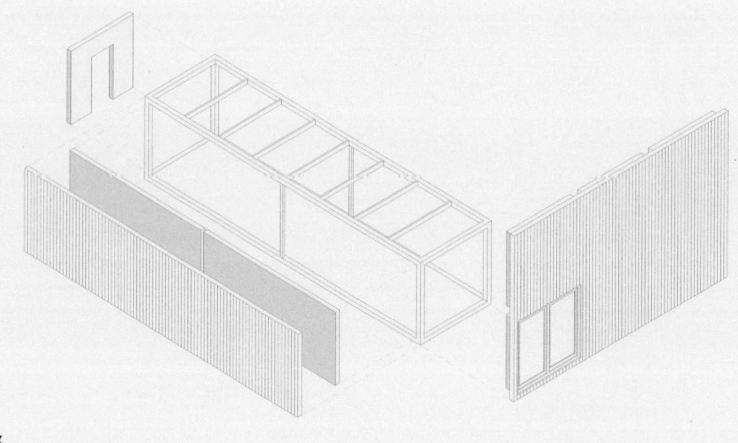

003

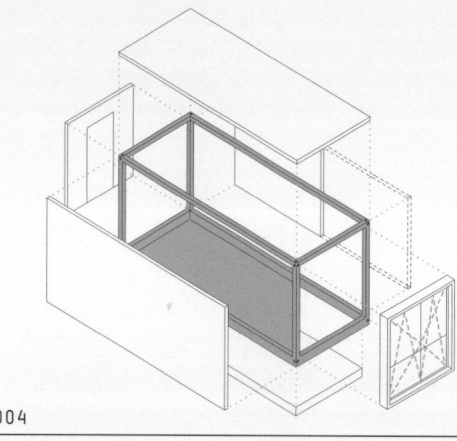

004

be industrially manufactured. In the case of mounting, disassembly and remounting where the components are reused directly without any modifications, this type of building system that is *open to the market* can be regarded as highly sustainable. Recycling of material or of products—for example in the case of the construction of a skeleton that has to be removed, prepared, or disposed of—is not necessary here. In addition, a supporting frame system can also accommodate do-it-yourself fittings, as the fittings do not have a load-bearing function, thus giving the project team complete design freedom in this regard.

003 | Module frame system
004 | Container frame system (with thermally separated frame profiles)

Building using containers has by now acquired something of a "cult status". The specific image of containers is often important when they are used in architecture, particularly in the event sector. The globalized container box evokes certain associations: the image of a well-traveled item is automatically linked with the raw atmosphere of a busy port, for example. The more striking the building solutions using (freight) containers, the stronger the associative effect of these buildings appears to be.

Despite their strict layout patterns, modular systems based on containers still offer a wide range of spatial solutions. The prerequisite for this is the positioning of the container—an originally very mobile and extremely unaesthetic box that is available everywhere and is always on the move; a quiet life at a fixed location is not something generally granted to a container. However, this does indeed happen when containers are used in architecture, but the container nonetheless remains mobile and transportable to a limited extent because it can be disassembled.

Building using containers thus often involves more than simply stacking and arranging containers in rows. An architectural structure can only be said to exist if:

- The mobile containers have a fixed location.
- Rooms and spatial connections/openings are created by architectural means, resulting in indoor rooms, intermediate areas, and outdoor spaces.

Only when containers are placed in a spatial context with spatial and architectural qualities do container boxes actually become *container architecture.*

# 03

USE OF CONTAINERS

Containers are always suitable for use where spatial solutions have to be found for a limited period of time. Container architecture is thus generally found in temporary buildings, where the advantages of a flexible, mobile building system come to the fore. The quick, short-term availability of containers make them effective as building modules. Unfortunately, minimal importance is often attached to architectural quality in the case of temporary building solutions. The frequent repetition of the same building blocks leads to the risk of architectural monotony and anonymity and urban-planning aspects are often ignored too. Construction based on containers used to have a negative image in the past, as a temporary requirement for space is often associated with circumstances where provisional interim solutions are required for people in emergency situations. This effect was often compounded by poor maintenance. However, current examples of container architecture show that solutions that fulfill high architectural standards are indeed possible using containers.

The use of containers in architecture is strongly influenced by the planned usage and the desired effect/impression. The usage strongly affects the type of solution implemented and thus the architectural design too, and low costs also play an important role. The planned service life, which is of course closely related to the usage, is another critical factor. The container is then no longer a mass product when used as a building module, but instead appears as something individual, unique and unmistakable.

The examples of container architecture allow for a categorization according to usage that emphasizes the profile and range of uses for containers in architecture. One can classify projects into public buildings, office buildings, and temporary housing and extensions to housing. Usage as permanent living space is financially most viable in locations where a mild, dry climate is present and protection against heat losses and moisture is not so important. For buildings that meet short-term spatial requirements in a functional manner, architectural design plays a less important role in the selection of the building module. The associated effect, that is attached to freight containers in particular, is used to help generate an image in the commercial sector, for events, for installations in public space, and for art projects involving containers: the image of the container then becomes associated with the product or event too, a process referred to as corporate architecture. Social projects also take advantage of the low acquisition costs of containers (low-budget or no-budget architecture).

A special form of container architecture involves recreated containers that are built using conventional construction techniques, but have the esthetic and structural characteristics of a building container or building container system—the so-called container-look. These are architectural quotes from a construction style that merely suggests a temporary character. In fact, this form of "imitation container" is less efficient from an economic and civil engineering point of view as the advantages associated with prefabrication are lost.

Use of containers

Public buildings
flyport is a passenger terminal that can be implemented anywhere in the world. It has a very short construction time, allows for flexible usage and design, and can be adapted to meet individual needs.

005 | flyport / wolfgang latzel architekten, 2004: Public buildings

Office
The headquarters of the PLATOON agency, a creative collective active in the area of communications, is located on an undeveloped site right in the middle of Berlin's lively Mitte district. The building consists of an ensemble of freight containers in military green, together with a lawn area and a pool.

006 | PLATOON headquarters / PLATOON, 2007: Office

Housing
This home in California uses freight containers as additional room modules. Used freight containers are freely available and thus also inexpensive in the USA because of the skewed balance of trade where imports exceed exports (Redondo Beach House / DeMaria Design).

007 | Redondo Beach House / DeMaria Design, 2008: Housing

Social/low budget architecture
Containers are suitable for use in charitable organizations funded by donations such as Children's Activity Centre (Phooey Architecture) in Melbourne, a social facility for children that doesn't charge for admission and therefore has to survive on public subventions.

008 | Children's Activity Centre / Phooey Architecture, 2007: Social low-budget project

Commercial/corporate architecture
The mobile PUMA salesroom by the architects LOT-EK consists of 24 shipping containers staggered to form a three-story sculpture. The product being sold, sports and leisure footwear will of course already have traveled the globe in just such a container. The associative interplay of product and architecture results in "corporate architecture".

009 | PUMA / LOT-EK, 2006: Corporate architecture

Event/exhibition
The Nomadic Museum (Shigeru Ban) employs containers in two ways: as building blocks that form the supporting structure for this large-scale exhibition hall, and as transport containers for building elements and for the exhibition display specimens themselves. Additional containers can also be rented locally, as required.

010 | Nomadic Museum / Shigeru Ban, 2005: Exhibition

The Illy Café (Adam Kalkin) was developed for a temporary use with a predefined duration as part of the 52th Biennale in 2007 in Venice. At the push of a button, a container that appears unmodified from the outside folds out into a café platform, thus becoming an exhibition object itself.

011 | Push Button House – Illy Café / Adam Kalkin, 2005: Event

Art
The Belgian architect and artist Luc Deleu uses freight containers in his art by stacking them to create monumental structures and landmarks that can be seen from far around. The containers used are often exaggerated in an artificial manner because of the way they are stacked without their essential forms being changed.

012 | Middelheim Construction X / Luc Deleu, 2003: Art

Container look
The projects in the "Container look" category are built in a conventional way, but are intended to resemble containers. These buildings merely refer to the esthetic and structural features of containers or container systems, and actually have little in common with container architecture.

They are often structural elements of secondary importance, such as additions to roofs or extensions (parasite architecture). The addition of these "pseudo-containers" is often a deliberate attempt to suggest a process of subsequent, often temporary extensions to a building, even when these secondary components were actually built at the same time as the main building structure itself.

013 | Sjakket Youth Center / PLOT = JDS + BIG, 2007: Container look
014 | Wismar Technology and Business Center / Jean Nouvel with Zibell+Partner, 2003: Container look

The use of imitation containers is somewhat odd from an architectural point of view, and this approach is also less than favorable from a civil engineering point of view since the imitation of prefabrication using conventional construction methods is actually very inefficient.

Nonetheless, an imitation container can help to liven up the architectural impression made by a building.

It is possible that container-like spatial cells are recreated for practical reasons in a certain context instead of using ready-made containers (for example: the facade openings might be too small to allow containers to be transported into the building). One possible alternative here would be to use special constructions that can be collapsed (e.g. folding containers).

Certain buildings appear similar to container architecture based on their architectural structure, but these are often simply skeleton structures that follow a strict modular layout. The appearance of a building constructed from spatial cells is created by borrowing the proportions of containers for the pattern dimensions and by repeating identical openings at regular intervals.

015 | Distributiecentrum Piet Zoomers / Van den Belt & Partners, 1992: Container look
016 | Student housing / Mecanoo Architekten, 2009: Container look

Use of containers

005

006

007

008

009

010

011

012

013

014

015

016

04
FUNDAMENTALS
CONTAINERS AS BUILDING MODULES

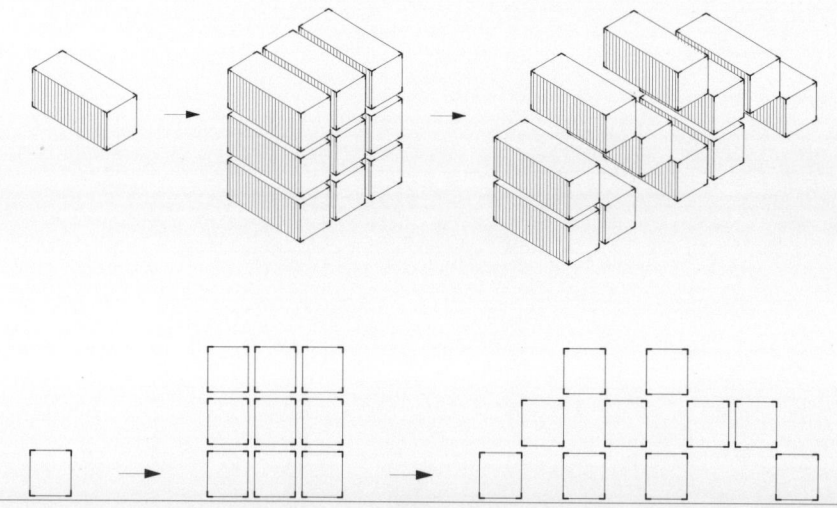

017

DEFINITION

Containers were originally conceived as vessels for goods, but are also used as spatial modules in architecture. The word "contain" comes from the Latin "continere", and means to hold together, to surround, to store. A container is a vessel that surrounds a usable volume of space and thus defines the spatial boundary between interior and exterior. As a walk-in box with useful spatial dimensions, the container fulfills the prerequisites for use as a spatial module.

Geometrically, a container is a six-sided rectanguloid with at least one opening. The standardized dimensions generally adhere to the internationally applicable ISO standard for freight containers. This standard has had a long-term influence on container use everywhere and provides the basis for the sizing of containers. These dimensions are primarily determined by the transport conditions and by the infrastructure present.[1]

The structure of a container is a combination of a frame and fillings. The frame supports the fillings that define the spatial boundaries; these fillings are generally also load-bearing and provide reinforcement for the building. Both static and dynamic loads must be taken into account when using containers. The static loads result from self-weight and the contents, while dynamic loads occur during transportation and mounting.

As a transportation vessel, the container is of course mobile. Transportation is of varying significance, depending on the usage in question. The importance of transportation when a container is used as a building module

Containers as building modules

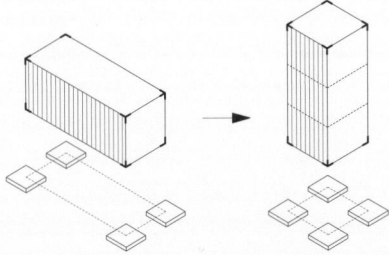

018

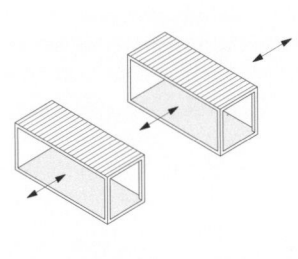

019

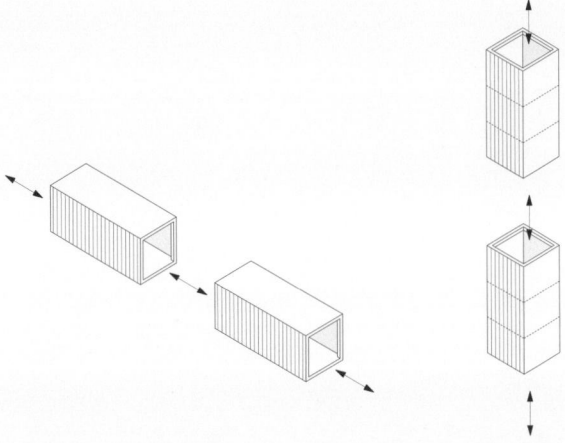

020

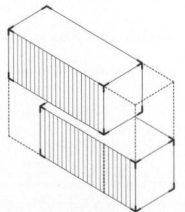 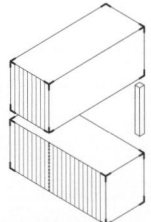

021

is different from that when it is employed as a shipping container: while a shipping container fulfills its function by transporting its contents, a building container only serves its actual purpose when it is used as a building.

The container has transport devices at its eight corners that are used to distribute all loads and to combine modules using joining devices to create a supporting system. The containers fit on standardized platforms, such as the chassis of a truck, where they can be fastened for transportation by road, rail or ship. The containers are moved using cranes and other lifting equipment.

PRINCIPLES

Containers are conceived as single-room modules that can form a flexible multi-room system once they are combined. The spatial system adheres to the modular layout system of container modules, which allows for relatively little freedom. Variations are possible within this layout principle, and empty and intermediate rooms can be created by omitting containers. However, it should be taken into account that vertical loads are distributed exclusively through the corners. When modules are combined, horizontal loads are also supported and transmitted by the joint devices.

017 | Containers as a single-room/ flexible multi-room systems (system-compatible)

Containers are generally positioned horizontally and supported at four points; they are designed for this scenario. Alternatively, containers can also be placed vertically to serve as symbols or landmarks—this represents a special case in terms of structural design, however. Experimental projects provide examples where containers are positioned vertically and are even designed as spaces with various accessible stories.

018 | Horizontal/vertical positioning

Interior connections between the containers are also horizontal/vertical, as required by the container layout. Because of load-bearing capacity constraints, the number of containers that can be stacked is between three and eight, depending on the construction system in question. To create spatial and functional linkages, appropriate openings are necessary that may be subject to certain constraints, depending on the system.

019 | Horizontal/vertical connection (longitudinal side)
020 | Horizontal/vertical connections (short side)

When combining container modules, it should be ensured that loads are supported by the corners, and thus that the corners are positioned above one another in order to transmit the loading to the ground effectively. If this is ensured, one speaks of system-compatible joints. In the case of deviations from this system, such as staggering or rotation of container modules relative to other modules, reinforcements are necessary at those points where corners meet a frame beam. These are then joints that are not system-compatible, which generally also result in a deviation from the standard dimensioning system.

021 | Staggering/rotation of containers (deviating from system)

When building using room modules, the system of dimensions of the room modules determines the axis grid pattern. Since a tolerance between the containers must also be taken into account, the axis grid length is made up of the container dimension plus the tolerance. This system also holds in the vertical direction. This interrelationship is also part of the dimension system in the ISO standard.

Containers as building modules

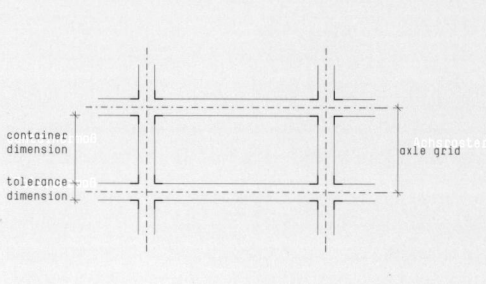

022

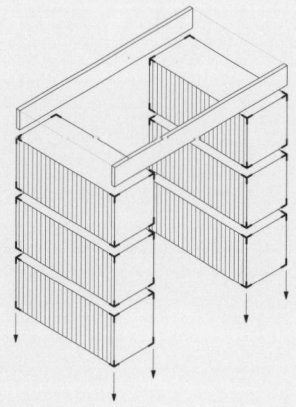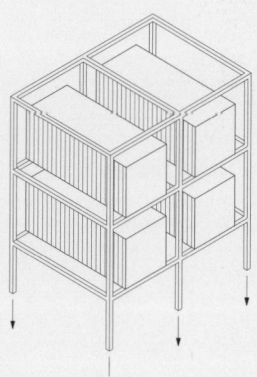

023

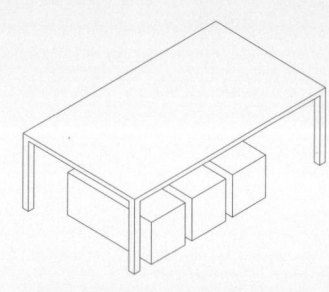

024

022 | Dimensional interrelationships when joining containers (axis grid = container dimension + tolerance)

In principle, containers can be joined to create self-supporting buildings where the containers act as the room-defining elements. However, it is also possible to use containers solely as a supporting structure, to stack them to form wall structures, and to create intermediate spaces using for example, a secondary roof-supporting structure that rests on wall structures made of containers. The structural strength of the container type is of critical importance here. Non-load-bearing containers can also be slid into a primary supporting structure and can thus define rooms without directly supporting loads from the rest of the building. This system provides maximum flexibility with regard to interchangeability and extensions for applications where connections between the various rooms are not of primary importance.

023 | Containers as a supporting structure/non-load-bearing containers

In the case of simple container building systems, every individual module has its own structural weather and climate shell. As container systems have certain structural engineering weaknesses, it may be a good idea to place the containers under a protective roof in order to avoid problems caused by leaks in the roof cladding, for example. The most radical solution is to place the containers inside an additional shell that fulfills a weather-protection and insulation function, if necessary.

024 | Containers with weather/climate shell and protective roof

FOUNDATION

Container buildings always require a foundation in addition to their overground structures. The type of foundation to be used depends on the geometry and more importantly, on the planned service life of the building. In the case of mobile buildings, foundation types that are demountable and leave no traces are to be preferred ahead of foundations made from site-mixed concrete.

Foundations that use large prefabricated concrete slabs (such as those made by Stelcon) are suitable as they can be installed easily and quickly. They are available in standard sizes between 2×1 meters and 2×2 meters. The slab thickness is around 14 cm. The subsurface must be prepared accordingly before installation is carried out. The slabs are placed on a mineral mixture and a sand layer of coarse and fine sand that compensates for unevenness of the ground.

025 | Slab foundation (e.g. Stelcon)

Alternatively, there are single foundations or screw foundations that are screwed into the ground using a motor or a similar boring machine. These can be easily removed in the same way, leaving behind no traces, and are thus even reusable. Some foundation types are compressed with granulate.

026 | Screw foundation

Prefabricated pad foundations made of pre-cast concrete parts (pin foundation, removable pre-cast concrete foundation) are also suitable as foundations. A precast concrete part is fixed in the ground here using steel rods sunk into the ground at an angle.

027 | Pin foundation

In general, the container must be fixed to the ground at its four corners using adjustable steel footings whose height can be evened out; these footings transfer loads to the ground. The fastening holds the building in place and also protects it from being lifted up by wind loads.

Since container buildings are very light, it may be necessary to provide a counterweight in the foundation area or to use weights to load down certain parts of the building—for example in the case of very high buildings. Solid, site-mixed concrete foundations can be used here, but for more temporary usages it is a

025

026

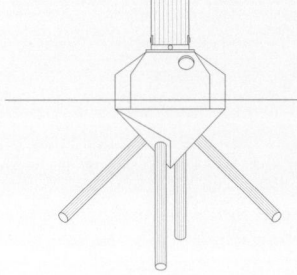

027

better idea to fill containers with sand or to use water tanks.

Floating constructions can also be implemented using containers. In this case, installation is either carried out on standard floating pontoons or on container pontoons that already have standard container dimensions (20 feet or 40 feet) and can be joined together. They are equipped with suitable corner brackets. The height of the pontoons is either the same as, one half of, or one third of the usual height of a standard container.

CONSTRUCTION PHYSICS

As a rule, civil engineering weaknesses in the structure of containers only allow for temporary usage. Containers can generally be classified as light building modules, which also has an effect on thermal insulation in summer and winter. The absence of solid matter means that container buildings have a low heat capacity. A change in the outdoor temperature thus leads to excessive cooling off or to overheating more quickly than would be the case with a solid structure.

Depending on the system, spatial modules are either insulated during production (building containers, module frames) or else have to be fitted with insulation later on (freight containers). They can then theoretically be equipped to meet almost any thermal insulation standard, but it should be remembered that large material thicknesses for interior insulation also result in a corresponding loss of indoor space. In addition, containers also have thermal bridges inherent in their design at their corners.

Conventional insulating materials are generally used. If there are higher requirements for thermal insulation in winter, or if it is important to make maximum use of the indoor space available, it can be a good idea to use vacuum insulation. This is more laborious to install, is more expensive and does not last as long (maximum manufacturers' warranties of 20 years) but has much better performance than standard insulating materials.

Spray insulation foams (blown-on insulation) are a space-saving solution, but can only improve the insulation capacity of a wall slightly due to their small coating thickness.

Alongside criteria regarding the structural stability of the supporting structure, fire protection is another important issue that must be taken into account at an early planning stage when constructing a building using containers. Building regulations generally provide very strict guidelines that should be checked in all individual cases and agreed on with the relevant fire department.

Even though steel does not burn (DIN building material class: A1) steel constructions generally have very low resistance to fire (F0). Because of their good thermal conductivity, steel structures suffer a loss of strength at the high temperatures that occur during a fire; this can happen at temperatures as low as 500 °C (in comparison, concrete is stable at up to 1,000 °C while aluminum looses its strength at around 300–350 °C). Additional measures are thus required in order to meet fire protection requirements. For example, the fire resistance of components can be increased by adding gypsum plasterboards or plaster coatings that foam up if a fire occurs (F30). Depending on the fire resistance of the structure, there may be approval limits with regard to multi-story designs.

CONSTRUCTION LAW

Construction law differs from country to country and state to state, meaning that it is difficult to make universally applicable statements regarding legal construction requirements. Further information about legal building regulations should be obtained from the relevant local authorities.

The status of temporary container constructions regarding building and finance law can be difficult to define. Container facilities are considered to be temporary[2] in Germany if they are installed for a very short period of time. This applies to construction-site containers for example, which are not classified as buildings; there is no need to apply for planning permission for these structures.[3] In all other cases, container buildings are subject to the same permission requirements as conventional buildings and do not have a special status. The applicable heat insulation ordinance must be adhered to in order to obtain permanent building approval (installation for longer than three years). A lower thermal insulation standard can be accepted for limited-period building approval.

ECONOMIC ASPECTS

Container facilities are generally erected with a significantly shorter construction period than would be necessary for conventional buildings. Short planning and implementation periods can reduce investment costs, and this advantage may tip the balance in favor of a container building.

Depending on the standard of fittings to be used, lower construction costs may also apply: industrial processes can produce container buildings more cheaply than would be the case with conventional construction methods. The costs should be considered relative to the planned service life of the structure. However, it should also be remembered here that additional costs for transportation, foundations and connections to utilities will arise. Regarding transportation, it should be taken into account that the worldwide shipping of a freight container is relatively inexpensive[4], even over great distances; however, the transportation of a building container by truck can be considerably more costly.[5] In particular, the transportation of special loads by road is laborious and expensive.

Prices for raw materials such as steel fluctuate strongly depending on demand on the world market. Possible future uses, resale of the modules after the planned service life, and disposal all influence the costs later on.

The required standard of fittings for the building plays an important role in the decision in favor of or against a container building, as it may only be possible to meet demanding civil engineering requirements in terms of insulation, fire protection, noise transmittal and other safety requirements to a limited degree. These issues should be examined for each project and the costs should be compared with the investment costs for a permanent structure.

Tax issues may also favor the use of container buildings. However, care should be exercised here: a structure made from containers is not necessarily allowed to be depreciated over a shorter period in Germany. This only applies to buildings that actually have a very short use period. The critical issue here is the degree to which a building is tied to its location. Evaluation of this criterion is independent of the potential mobility that is always inherent in container buildings. A container structure can be categorized as having a fixed location if a foundation base exists (which need not necessarily be permanently installed). Use as a permanent residence, the existence of sanitary facilities or connections to the public supply grid can all make a container building an integral part of the site from a legal evaluation point of view. Container buildings that are adjudged to be tied to their locations are classified as light buildings, and these can then be depreciated over a shorter period of time than solid-construction buildings.[6] Modular construction generally results in fixed-location buildings. Tax law treats these buildings in the same way as conventional buildings.

The decision whether to rent or buy the containers should always be based on the planned service life of the building and on any conversion work that may be necessary. Buying is generally a better option for usage periods longer than 24–36 months. The price for purchasing a new container module varies depending on the system, and used containers can also be bought as an alternative for around 50% of the new price.

ECOLOGICAL ASPECTS

Container architecture has significant advantages compared to conventional construction methods from an environmental point of view. As container building blocks are inherently demountable and remountable, it is possible to reuse the modules once the planned service life of the building is over. The building can be disassembled into individual, stand-alone spatial units. Modularity also means that the system can be extended.

Because of the short time it takes to erect a container building, it is possible to react more quickly and flexibly to increasing spatial requirements and to planning changes by gradually extending the structure. In addition, gradual building extensions are also possible. For example, a smaller part of a commercial building can be built initially for a company start-up, and the building then completed later on when extensions become necessary. In this way, unoccupied building space can be avoided.

The period of use of the container should be carefully considered. In principle, the expected service life of the basic frame structure (supporting construction) is longer than that of the fittings and finishing. The building services technology in particular, quickly becomes obsolete after a few years (technical equipment). The container is then disposed of after its full service life is over. The steel can be reused as a raw material, as it can be better recycled than concrete for example, which can only be reprocessed for lower-grade uses. The investor can even recoup a small portion of the investment costs by selling the recycled steel; other construction waste has to be disposed of at a cost, however. Depending on the particular design, a container can contain between 0.5 and 4.0 metric tons of steel, which can be sold at the scrap price[7] that is currently applicable. In the case of rental, the owner has to bear the disposal costs.

In addition, it is also possible to supply container buildings using alternative energy sources. For example, it is possible to integrate solar panels or photovoltaic equipment into the outer shell or the roof. These generate energy and also help to reduce problems with excessive temperatures inside the container building in summer by providing shading.

INTERNATIONAL ASPECTS

When comparing the international situation, it can be observed that containers are available and used as building modules the world over. However, it can be assumed that the type and frequency of use, the selection of container type, and the intended purpose are all strongly dependent on certain local considerations:

1) Climate: The requirements placed on the quality of the climate shell are as varied as the climates themselves. The heat insulation requirements are legally stipulated in some countries, while this is not the case in other regions. Because of the mild climate conditions in California for example, there are many residential projects based on shipping containers there that only have to fulfill modest requirements in terms of thermal insulation.

2) Distance: Long transportation distances and high costs for transportation will apply for regions with lower population densities. For this reason, the transportability of the building can be of much greater significance here as compared to other regions. One of the reasons the "trailer" model is common in the United States is probably because it is already fitted with a chassis and can thus be transported more easily and cheaply.

3) Cultural influences: Cultural norms can also play a role in the decision for or against a temporary (container) building. Mobility is an integral part of people's living habits in the USA, for example. When moving home in rural areas, it is sometimes even possible to take advantage of light modular constructions and effectively take your house with you. In Germany, buying real estate and private homes is still seen as a long-term investment, meaning that more solid constructions tend to be favored. The image associated with container buildings varies too from region to region: although a lot of construction is taking place in China, container buildings are not widespread because container units with trapezoidal sheeting are regarded as inferior.

4) Global economic influences: Economic conditions can also influence the use of containers; in particular, the type and amount of trade (import/export) plays an important role here. For example, a lot more goods are imported into the USA in shipping containers than are exported out of the country, meaning that there is an excess of containers at American container terminals. Thus, they are sold relatively cheaply. In addition, the financial crisis that began in 2008 has also led to an excess of freight containers worldwide simply because less goods are being transported than before. The currently falling prices for steel are also reducing prices and production costs for containers.

5) Regional economic circumstances: Cost-effective building can be a significant factor in economically less-developed regions. Commercial and educational buildings and infrastructural facilities such as airports can be built for low investment costs in order to stimulate the economic development of certain regions. In addition, a building system that can be erected at short notice can generally react a lot quicker to changing economic circumstances.

6) Structural statics requirements: The requirements for structural stability can also vary depending on the region in question. Steel constructions are particularly suitable in areas with a risk of earthquakes such as Japan, as they exhibit elastic behavior when subject to vibrations.

# 05

CONTAINERS AS BUILDING BLOCKS

Containers are universally applicable building modules in the construction sector. They are available in various types and for various purposes. The range of uses is just as wide: from functional buildings right through to experimental, architecturally demanding custom solutions.

The following categorization forms the basis for the design consideration of container architecture—containers will be considered under the headings: freight containers, building containers and module frames/container frames.

028 | Isometric drawing of a freight container

TYPE 1
FREIGHT CONTAINERS

For a number of years now, freight containers have been subject to reappropriations in terms of their function and have been finding increasing use in architecture. There are two main aspects that make containers attractive in architectural applications: their characteristic design that can be used to create a certain image, and their system-based advantages such as prefabrication, mobility, modularity and global availability.

However, these benefits must be weighed up against certain civil engineering disadvantages too. Containers are accessible when loading and unloading their contents, and offer a sealed shell and a certain minimum amount of weather-protection depending on the goods being transported; however, they cannot fulfill important civil engineering requirements for a building shell such as provision of daylight and protection against heat, moisture, noise and fire unless certain conversion and extension measures are carried out.

The space inside a container is only suitable for human occupation to a certain extent because of the limited height on offer. The height is sufficient to be able to stand or walk upright, but more comfortable room heights, which may also be stipulated by construction regulations, can only be achieved with high cube containers.

Nonetheless, the appropriate use of freight containers in construction can lead to quite remarkable architectural solutions. These are characterized by an intelligent architectural treatment of the properties and features of containers. There are three main strategies that can be taken:

<u>Acceptance:</u> The freight container is used in its original form, and the existing finishing and product characteristics are accepted as part of the architectural design. The usage concept is focused on the container not as a place where people will spend a lot of time, but rather as a pre-defined room with a correspondingly low standard in terms of fittings and equipment. In certain cases, it may not even be possible to enter the container, which is instead used as a supporting structure, has advertising mounted on it, or is itself part of an artwork.

029 | Speybank / Luc Deleu, Yokohama, Japan, 2005

029

030

<u>Fittings and equipment:</u> A freight container is treated as a raw building module that can be fitted with further equipment and features. The container can then be converted and upgraded subsequently to the desired standard with regard to fittings and technical equipment. This type of use can be particularly labor-intensive when constructing residences because of the high standard of fittings required. However, the advantages of a high degree of prefabrication can also be exploited by carrying out the conversion and upgrading work in a workshop.

030 | Campus / Professor Han Slawik, Almere, the Netherlands, 1992

<u>Combination:</u> To bring the fittings standard to an acceptable level, the container is combined with further components that fulfill the requirements of the planned building usage. This could include an external facade or roof construction, or a completely separate climate shell (house inside a house). A cost-effective alternative is the use of containers indoors only—for example, to divide off smaller functional units inside of larger, existing building

structures. In this case, no labor-intensive conversion or fittings work on the container is necessary; instead, minimal conversion measures suffice to make the container suitable for use as an occupied space.

031 | 1001 days/Meyer en van Schooten Office, Amsterdam, the Netherlands, 2000

DEFINITION

The freight container system cannot be defined solely by the properties of the individual containers. Containers are the core of the system, but only form a transport system when other aspects are considered such as logistics, transport and their ability to be combined.

This is emphasized by the main requirements placed on a uniform transport system, which always has the aim of simplifying and increasing the efficiency of the shipping of goods:
- Uniform transport of differing loads using uniform containers
- Closed transport chains with no reloading of goods along the way
- Reusable containers

The implementation of this sort of transport system called for the specification and standardization of its main elements, and this is exactly what was achieved with the introduction of the ISO standard internationally. This standard is updated at regular intervals and is adjusted for technical and logistical conditions.[8]

STANDARD

The International Standards Organization (ISO) established a committee with 26 participating countries in 1961 with the goal of standardizing the transport systems that until then, had been very different in Europe and America.

The main aim of the ISO standard that was eventually approved by the 26 participating countries in 1964, was to ensure the functionality and compatibility of the container transport system. For this reason, the standard mainly deals with dimensional, geometrical and use-dependent specifications. On the other hand, the production and structural design of containers remain largely untouched by the standard.

America's influence led to all container dimensions in the ISO standard being specified using the imperial measurement unit of the foot (conversion: 1 foot = 0.30479 meters). This meant that the specification of the interior dimension of the 8-foot-wide ISO container did not allow for loading with two adjacent Euro pallets (80×120 centimeters), with the result that the width is a few centimeters too narrow for this purpose.[9]

The core of the container standard is ISO 668, which contains all specifications necessary for the standardization of the container system. Alongside the definition and classification of standard containers, the standard also deals with all relevant dimensions (internal and external dimensions, door opening dimensions) the construction tolerances for these, and the geometry and positions of the corner fittings for transportation and fastening. Freight containers that are manufactured to meet the specifications of ISO 668 are referred to as ISO containers.

Further standards deal with the terminology of various container types (ISO 830), classification (ISO 6346) and the specification of corner fittings (ISO 1161).

DIMENSIONS

To ensure optimum compatibility, the dimensions of standard containers are modularly structured. The starting module here is the 40-foot (type A) long container, and the other sizes (nominal sizes: 10 feet [type D], 20 feet [type C], 30 feet [type B]) are fractions of this starting length. In order to facilitate stacking and moving of the containers, low tolerances of ¼ inch per 10 feet are taken into account, meaning that a nominal 20-foot container is actually only 19 feet 11½ inches (6,058 mm) long.

032 | Principle of the ISO dimension system
033 | Dimension grid of freight containers as per the ISO standard
034 | Table of container dimensions

The 20-foot container is the most widespread size. It has been adopted for the TEU (twenty-feet equivalent unit) measurement unit that is used in specifying transportation and manufacturing capacities. 20-foot and 40-foot containers have established themselves in shipping and truck transportation because of the ease with which they can be combined.

A standard width of 8 feet has been specified for freight containers to suit the requirements of road transport. Container widths that deviate from this can only be integrated into the ISO system if the positions of the corner fittings remain unchanged.

The standard height (standard cube) is 8 feet 6 inches. In addition, there is the low cube (height 8 feet), and a high cube that was also introduced into the standard in 1992 (height 9 feet 6 inches). The height dimensions are not modular, but are instead based on structural considerations, as the containers can be stacked individually regardless of their height.[10]

035 | High cube beside standard cubes

Alongside the standardized dimensions specified in the standard, containers with special sizes are also developed every so often,

031

035

036

Containers as building blocks

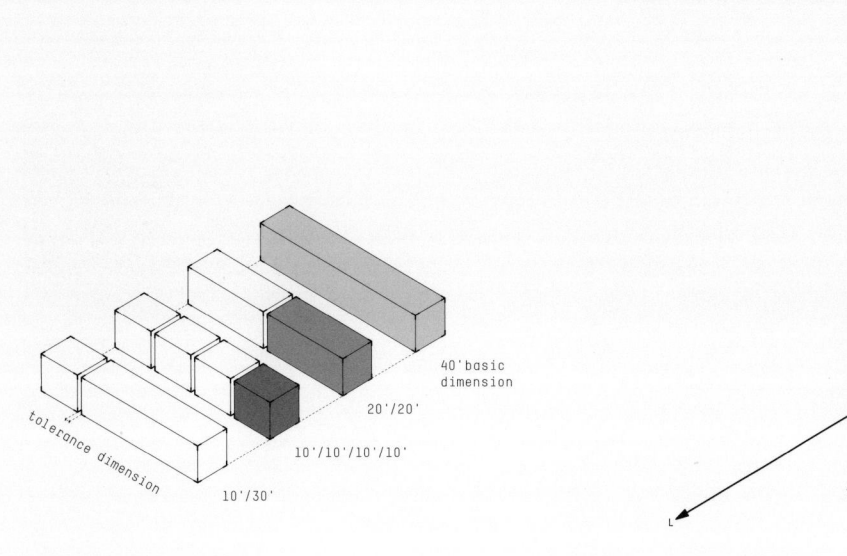

032

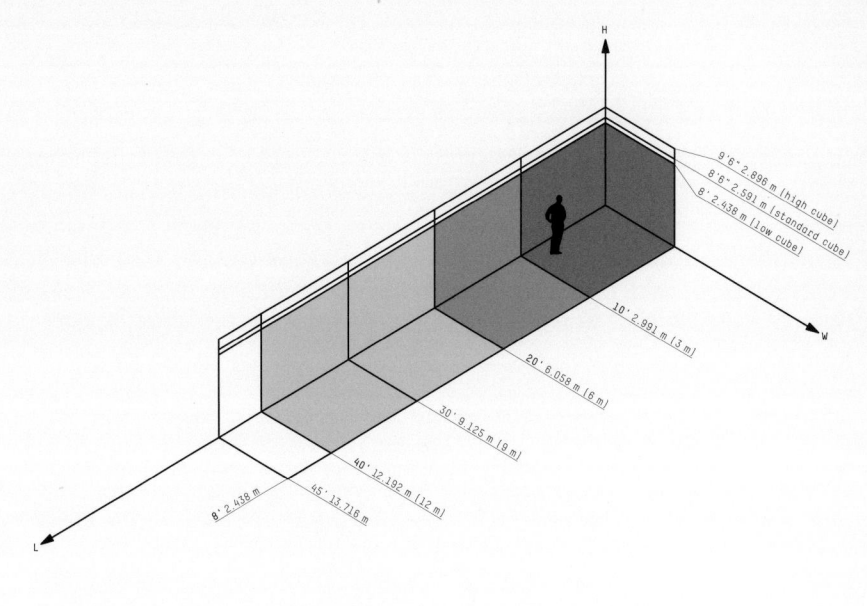

033

Desig- nation	Nominal size	Nominal height	External length		External width		External height		Minimum internal dimensions			Door opening dimensions	
									heigth	width	length	height	width
1AAA	40'/12m	High cube	40'	12,192 mm	8'	2,438 mm	9' 6"	2,896 mm	2,655 mm	2,330 mm	11,998 mm	2,566 mm	2,286 mm
1AA	40'/12m	Standard cube	40'	12,192 mm	8'	2,438 mm	8' 6"	2,591 mm	2,350 mm	2,330 mm	11,998 mm	2,261 mm	2,286 mm
1A	40'/12m	Low cube	40'	12,192 mm	8'	2,438 mm	8'	2,438 mm	2,197 mm	2,330 mm	11,998 mm	2,134 mm	2,286 mm
1AX	40'/12m	—	40'	12,192 mm	8'	2,438 mm	<8'	<2,438 mm	—	2,330 mm	11,998 mm	—	2,286 mm
1BBB	30'/9m	High cube	29' 11¼"	9,125 mm	8'	2,438 mm	9' 6"	2,896 mm	2,655 mm	2,330 mm	8,931 mm	2,566 mm	2,286 mm
1BB	30'/9m	Standard cube	29' 11¼"	9,125 mm	8'	2,438 mm	8' 6"	2,591 mm	2,350 mm	2,330 mm	8,931 mm	2,261 mm	2,286 mm
1B	30'/9m	Low cube	29' 11¼"	9,125 mm	8'	2,438 mm	8'	2,438 mm	2,197 mm	2,330 mm	8,931 mm	2,134 mm	2,286 mm
1BX	30'/9m	—	29' 11¼"	9,125 mm	8'	2,438 mm	<8'	<2,438 mm	—	2,330 mm	8,931 mm	—	2,286 mm
1CC	20'/6m	Standard cube	19' 11½"	6,058 mm	8'	2,438 mm	8' 6"	2,591 mm	2,350 mm	2,330 mm	5,867 mm	2,261 mm	2,286 mm
1C	20'/6m	Low cube	19' 11½"	6,058 mm	8'	2,438 mm	8'	2,438 mm	2,197 mm	2,330 mm	5,867 mm	2,134 mm	2,286 mm
1CX	20'/6m	—	19' 11½"	6,058 mm	8'	2,438 mm	<8'	<2,438 mm	—	2,330 mm	5,867 mm	—	2,286 mm
1D	10'/3m	Low cube	9' 9¾"	2,991 mm	8'	2,438 mm	8'	2,438 mm	2,197 mm	2,330 mm	2,802 mm	2,134 mm	2,286 mm
1DX	10'/3m	—	9' 9¾"	2,991 mm	8'	2,438 mm	<8'	<2,438 mm	—	2,330 mm	2,802 mm	—	2,286 mm

034

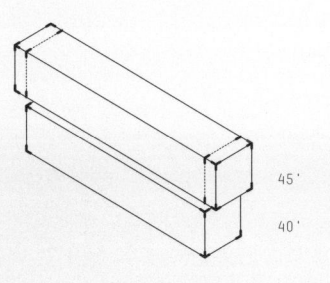

037

mainly with the goal of maximizing economic efficiency. The main special lengths in use for containers are 45 feet, 48 feet, 49 feet and 53 feet, as these can be integrated into the existing module system by employing additional corner fittings. A 45-foot container can be transported by road on a standard chassis.

036 | 45-foot containers stacked with ISO containers
037 | 45-foot container

Completely new length module systems are also being discussed, such as a 49-foot/24-foot module that could replace the existing 40-foot/20-foot module in the long term.

The so-called "pallet wide" containers developed for the efficient loading of Euro pallets are compatible with the ISO standard despite having internal dimensions which are a few centimeters larger, as the frame construction including corners is unchanged even though the external walls have been shifted relative to the standard container.

The DB inland containers that are solely intended for use by German State Railways on its freight trains are no longer in conformance with ISO with their width of approximately 2.5 meters, but were designed to make optimal use of the rail gauge of the German railway network and are thus an island solution providing the most efficient solution for rail freight.[11]

WEIGHTS AND LOADS

Freight containers are designed to be able to carry heavy loads, and can thus support a

Containers as building blocks

Container	Self-weight	Loading	Total weight
45'	5,000 kg	25,680 kg	30,480 kg
40'	4,000 kg	26,480 kg	30,480 kg
30'			25,400 kg
20'	2,330 kg	21,670 kg	24,000 kg
10'			10,160 kg

038

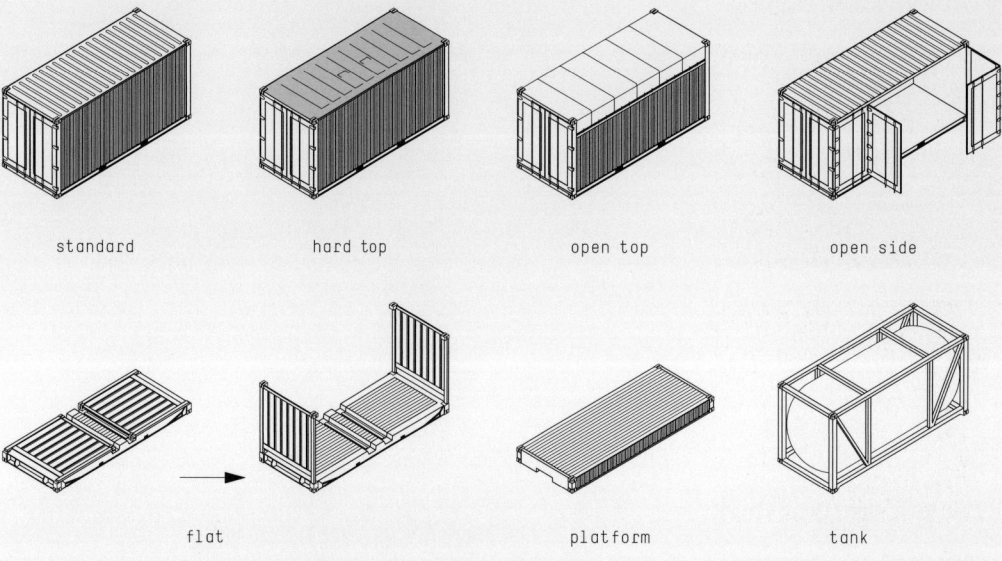

039

038 | Table with self-weights and loadings

multiple of their own self-weight—for example, a 20-foot container can support ten times its own weight. Containers are thus generally considerably oversized for use in architecture.

Self-weight is an important criterion with regard to the quality of a container. The minimum weight must also be observed, as it reflects the amount of steel used and thus indicates that the necessary static stability has been provided.

The maximum loading of containers is generally not determined by the permitted loading weight but instead by the loading volume, as the permitted total weight is generally only achieved with particularly heavy loads. For this reason, comparatively low loadings are accepted for long containers (40-foot, 45-foot), as the permitted loadings for the corner fittings, which are the same for all containers, must not be exceeded.

TYPES

Based on the standardized dimensions, there is a variety of series container types that are mostly defined by the ISO 830 standard (container vocabulary).

The various types have been developed based on their suitability for the load in question so that different types of goods (piece goods/bulk materials, perishable, heavy, bulky or liquid loads) can be transported in a uniform manner.

039 | Freight container types

A standard container has five closed sides and an opening at one end with a double-leaf door. It is intended for piece goods of all kinds.

040 | Standard containers

The hard top container can be loaded with heavy, bulky goods thanks to its completely removable roof. The container frame is designed to be accordingly stronger.

An open top container is open at the top, but does not have a fixed roof. The cross beams above the doors can be removed to improve loading flexibility. The open roof is generally covered with a plastic sheet.

041 | Stacked, covered open top containers

The longitudinal side of open side containers can be opened completely. The necessary

040

041

042

043

reinforcing bars can be removed to facilitate loading by forklift, for example.

A bulk container has filling openings in the roof and emptying devices that are generally located in the door leaves.

Platforms have a permitted total weight of 34 metric tons and are designed for particularly heavy loads. They consist solely of extremely strong base constructions that are higher than those on standard containers. They can be stacked when not loaded, and can be transported on conventional chassis.

Container flats consist of a high-strength floor construction with two foldable end walls. These end walls make it possible to stack a number of loaded flats on top of one another, which is not possible with platforms.

042 | Locking for flats (conical)

A ventilated container, which is aerated through splash-proof ventilation openings in the top and bottom cross beams, was originally specially developed for the transportation of raw coffee, which is the reason why they are also referred to as "coffee containers".[12]

Thermal containers have an insulated outer shell that prevents or at least minimizes temperature fluctuations inside the shell. When fitted with the appropriate equipment, this container can also be used as a refrigerated container (or "reefer container").

A tank container consists of a strengthened and reinforced container frame with an integrated pressure tank where chemicals, fuels or foodstuffs such as fruit juices or liquor can be transported.[13]

In addition to these standardized container types, there are other special designs for specific purposes that could also come into consideration for architectural uses.

For example, there are containers with openings that vary in number, size and position. Containers with doors at both ends are designed so that they can be loaded at both ends, but they could also be glazed at both ends (to provide daylight) as the frame construction allows for this. Containers with openings in the side walls allow for containers to be combined to create larger spatial units.

043 | Containers with side doors and openings for loading

Containers can be equipped with additional fittings to make them suitable for certain types of loads—for example, with textile linings (so-called "liner bags") or other special technical equipment.

COSTS

In principle, two different quality levels can be identified in the international container trade:

Used containers
There are no standard criteria for used containers because of the different uses and loads they may have had. They are thus generally bought only after visual inspection. They are a number of years old, show signs of their age, and may be damaged or dirty.

New container (almost new/as new)
It is almost impossible to buy factory-new containers in Europe. "New" containers have generally already been used for at least one loaded sea journey from the Far East. However, the value of an individual container does not drop significantly because of this one sea journey. There will generally be little damage or signs of use.

The price of freight containers depends on the (often quite considerable) fluctuations in the price of steel and the exchange rate for the dollar. The price fluctuations are amplified by the purchasing policy of large shipping companies, which directly affects demand, supply and thus also the price. In addition, the costs for storage and logistics for unused containers, which are often considerable, are an important consideration for shipping companies.

Political crises can also influence availability, as the movement of troops and of military equipment also gives rise to enormous transportation requirements.

Another cost factor results from geographical factors relating to goods transport. For example, freight containers are less expensive in regions with an excess of imports (e.g. the west coast of the USA) than in areas where exporting dominates (e.g. Southeast Asia).

Over the course of 2008, prices for new containers rose by around €300 and those for used containers rose by around €150. The price (including delivery) for a new 20-foot container (TEU) is currently around €2,400–€2,600, and that of a used container is around €1,200–€1,400.[14]

The quality standard of the various shipping companies should be taken into account when buying used containers, as these companies often demand higher manufacturing standards so that they can save on maintenance work later on.

Rental costs for containers (TEU) vary considerably depending on the duration of rental, and delivery costs also play an important role. As a rule of thumb, rental charges of around €100 per month can be assumed, dropping to around €50 per month for longer rental periods. Thus the rental costs exceed the purchasing price after two years at the latest. Considering the residual value too, it is clear that rental is only viable for very short rental periods, as is the case with transport usage. Purchasing is the only realistic option for building applications, especially for conversion projects.

As delivery is a significant contributor to costs, efficient transportation solutions such as the delivery of a number of containers together should be used.

044 | Used, damaged container/rusted container

Converting a container into a useful building mainly involves manual work: metalwork and welding to create openings and reinforcements etc., carpentry and insulation for the interior fittings, sealing of the roof and plumbing. The costs of this work depend very much on the degree and type of the conversion, and also strongly vary regionally and nationally because of differing pay levels. The total conversion costs can quickly exceed the cost of purchasing the raw container.

The following costs can be assumed for standard tasks: installation of a standard door approx. €550 (installation alone), installation of a window approx. €480 (installation alone), 50-millimeters inner insulation with particleboard cladding approx. €1,800 (20 feet), separation wall approx. €200, exterior painting in a RAL color approx. €600 (20 feet). These costs only account for the necessary openings in the trapezoidal sheeting and the use of reinforcements. The components themselves (doors, windows etc.) are not included in these estimates.

045 | Conversion of a container/workshop made of old containers
046 | ISO container with side door and window/various container types stacked

TRANSPORTATION

Regardless of whether piece items, bulk or waste materials, perishable, stable or toxic goods are to be transported—the use of ISO containers turns all goods into standard loads that can be transported together as part of a closed logistics and shipping chain.

The most important means of transportation for containers is <u>transport by sea and waterway</u>. The largest container ship in the world is currently the "Emma Maersk", owned by the Danish shipping company Maersk; it is 397 meters long and has a capacity of up to 15,000 TEU.[15]

The development of container transportation over the last few years is characterized by continuous increases. Particularly noteworthy is the prominence of Asian ports, where container movements have almost doubled since 2002.[16]

047 | Table: Container movements at the 20 largest ports in TEU units

A significant fall in global container transport (approx. 12%) has been predicted for 2009 because of the worldwide financial crisis.[17] Goods movements at the port of Hamburg for the first nine months of the year fell by 35%, for example. This will result in more used containers becoming available, which can then be put to new, intelligent uses.

Road transportation by truck offers advantages in terms of the quick, flexible final distribution of goods and thus ideally supplements shipping by water.

A 40-foot container must be securely locked to the chassis using twistlocks before being transported in this way. Variants of the standard chassis can have a number of locking points (to secure two 20-foot containers), may have variable length, and may be able to tip the container or position it in a certain way.

A gooseneck chassis is a particularly flat chassis for high cube containers, which reduces the overall height of the trailer to a minimum.

048 | Container transport by truck

Container transport by rail is not as flexible as road transportation, but can be just as viable because of its loading capacity if the rail network is well developed and if suitable transshipment terminals are present. It is even possible to stack containers (two stories) for transport on freight trains, as is done in North America.

Test trains are now even traveling between Europe and Asia with the aim of eventually establishing a land alternative to shipping by sea. A German freight train from Hamburg to Beijing currently takes around 15 days, which is half as long as the same journey by sea.[18]

PRODUCTION

For a number of years now, the production of freight containers has been carried out almost solely in Asia—mainly in China, India and Indonesia. The reasons for this shift to production locations in Asia are the low wages and production costs and the proximity to Asian ports where exports dominate.

Containers are produced from prefabricated semi-finished products (rolled steel profiles, trapezoidal sheeting), which are then processed and assembled mostly by hand.

Because of the resulting fluctuations in quality, every container is generally subjected to various functional tests that check the stability, load-bearing capacity, watertightness and weight of the container.

Modern factories have a number of production lines. A production line can produce up to 70,000 units per annum, which corresponds to a frequency of around 8 minutes if uninterrupted production is assumed.

049 | Table: Production data for all Singamas Ltd. factories

044

Containers as building blocks

045

046

Year	2008	2007	2006	2005	2004	2003	2002
Total TEU	247,422,079	232,329,888	205,118,538	183,293,585	163,681,152	140,786,126	113,433,856

047

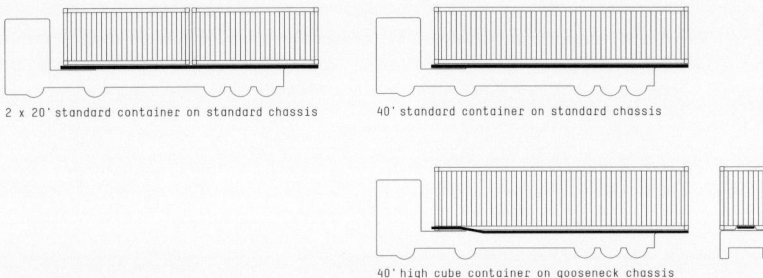

048

Year	2003	2004	2005	2006	2007	2008
Total TEU	466,523	618,836	494,282	583,543	838,638	567,400

049

CONSTRUCTION

Freight containers are subject to extreme mechanical, climatic and chemical conditions during their everyday use. Repeated loading and unloading, lifting and lowering, and transport exposure to wind and weather in saline environments all make significant demands on the properties of the material used to make the containers.

For this reason, almost all freight containers are made of COR-TEN steel nowadays, a steel alloy whose name is derived from the two main material properties of CORrosion resistance and TENsile strength.[19]

Containers made of other materials such as aluminum, wood and plastics are less common as they have disadvantages in service.

For example, aluminum containers are significantly more expensive and less stable than steel containers, and are also more difficult to repair and convert as they cannot be welded.

Plywood containers with non-load-bearing wooden walls are less expensive than steel containers, but are not as stable or as resistant to moisture. In addition, their frames must be more generously dimensioned as the cladding is not load-bearing.

A frame made of various steel profiles welded to each other is the primary supporting structure. Different profiles are used depending on the position and loading involved: angle profiles, C profiles, rectangular profiles and even special cross sections with a sheet thickness of up to 4.5 millimeters. A secondary carrier-member grid provides reinforcement for the container base. The frame constructions are standardized in terms of type structures, meaning that no additional structural calculations are necessary apart from the compulsory loading tests.

050 | Detail of the frame structure

The walls of the steel container are made of 2-millimeters-thick trapezoidal sheeting that is welded into the container frame all the way around, just like the roof that is made of creased steel sheeting. Alongside its room-defining function, the wall-filling surface stabilizes and strengthens the frame and also bears loads from the supporting structure.

Because standard containers have a wall structure made from steel sheeting, their un-ventilated version will be vapor-proof; this can

Containers as building blocks

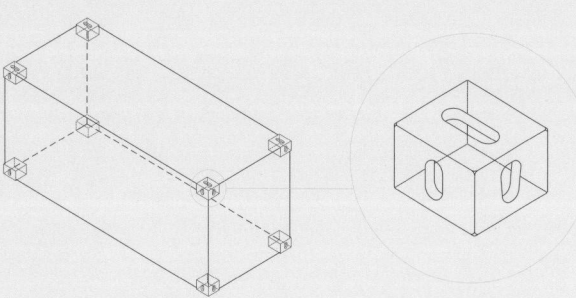

052

050

051

easily lead to the formation of condensate on the inside of the walls if the container is used as a building. This should be taken into account when converting containers into living quarters, as insulation is generally fitted inside the walls here.

So-called corner fittings are a common feature of freight containers. As the attachment points for loading, fastening and stacking, they are the basis of the overall transport system. For this reason, they are also dealt with by a separate standard (ISO 1161).

Corner fittings are made from cast steel and form the eight corners of each container. Every corner fitting is equipped with different openings on the three sides. The upper and lower openings, which are designed as slot holes to compensate for tolerances, are used for attachment when stacking containers; the side openings are mainly used for attaching the lifting and fastening equipment.

051 | Corner fittings
052 | Corner fittings

The base of a standard container consists of 28-millimeters-thick, 19-layer plywood sheets that are screwed to the cross-member grid. The sheet thickness can vary depending on the loading.

Plywood bound by phenol resin is used here because of its resistance to both pressure and moisture. However, these sheets are not 100% moisture-proof, meaning that contamination with chemicals or fuels and the resultant vapors later on cannot be excluded in the case of used containers.

The opening at the ends for loading purposes is closed off with a two-leaf sheet-covered steel frame door. The door leaves can be opened through 270°, meaning that they can be positioned parallel to the side walls. Each door leaf can be locked using two bolts and corresponding supporting devices.

The seals on the container doors are made of EPDM, a synthetic rubber that has good elasticity, weather and moisture resistance, ozone resistance, and good thermal and chemical durability.

The COR-TEN steel surface is extremely resistant to weather and corrosion. The iron oxide layer that forms on an untreated surface is not permanent, however. If additional surface coating is not carried out, particles continually detach themselves from the corrosion layer, which would seriously detract from the performance of the container.

For this reason, containers are given a highly durable, epoxy-based, three-layer corrosion and paint coating. Damage to the coating cannot be avoided during service, but is not harmful thanks to the high steel grade.

There are hardly any limitations with regard to the choice of color. Shipping companies or renters often use the containers for advertising purposes and add their own corporate design. The result is the familiar colorful appearance of container ports and ships.

TECHNICAL EQUIPMENT

Containers are fitted with additional technical equipment as required. For example, cooler containers have permanent or retrofitted coolers (clip-on-cooler) that are connected to the electricity supply during storage or transport by sea. They can also be operated using combustion motors while the containers are being transported

057

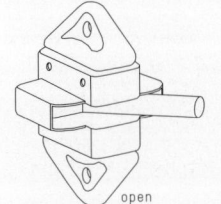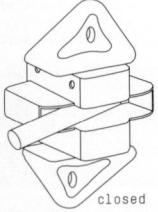

053

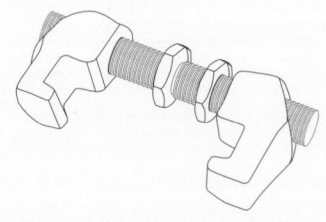

055

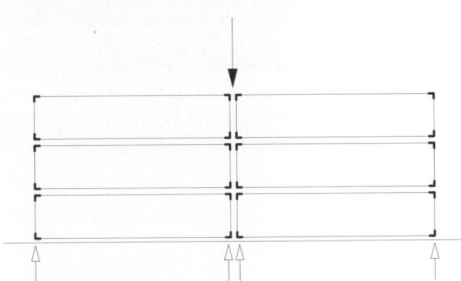

054

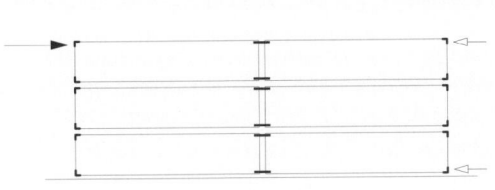

056

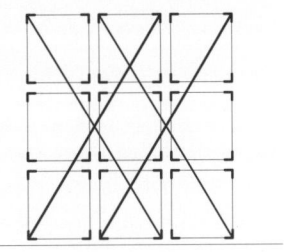

058

by land. Mechanically ventilated containers are supplied in a similar manner.

Other items of technical equipment are confined mainly to transportation and transshipment technology. Technical equipment for security and for locating containers (GPS) has become increasingly common in recent years. Containers are fitted with chips that record and process identification and other data regarding loadings, contents and routes. In addition, modern container terminals have comprehensive security equipment to monitor and identify containers from the outside.[20]

Transport-specific equipment is only relevant for building use in specific cases; conventional heating and electricity supplies can generally be retrofitted.

SECURITY

Containers are securely attached to each other and to the substructure using corner fittings to provide optimum security for loads. This also applies when containers are used in building applications.

So-called twistlocks are used to connect the containers vertically and to anchor them to the subsurface in question; the twistlocks are inserted into the slot holes in the corner fittings and locked by turning the lug through 90°. Semi-automatic and fully automatic locking twistlocks are now in use in sea transport, which lock and release when they click into the corresponding fittings on the containers already present.

053 | Twistlock
054 | Vertical load distribution

So-called bridge fittings are used for horizontal connections in the lateral and longitudinal directions, bridge fittings connect two containers to each other using a threaded rod at the corner fittings. Bridge fittings are also used to transmit horizontal building loads (wind forces).

055 | Bridge fitting
056 | Horizontal load distribution

Various fastening systems are used to securely fasten containers in place on container ships, depending on the requirements. Rail systems for wall attachment are used together with base fittings that fasten the bottom container to the floor of the ship. Freely standing stacks of containers are strapped and attached to each other using lashing systems.[21]

057 | Lashing system/detail shot of lashing procedure/container being lashed
058 | Diagonal reinforcement using lashings

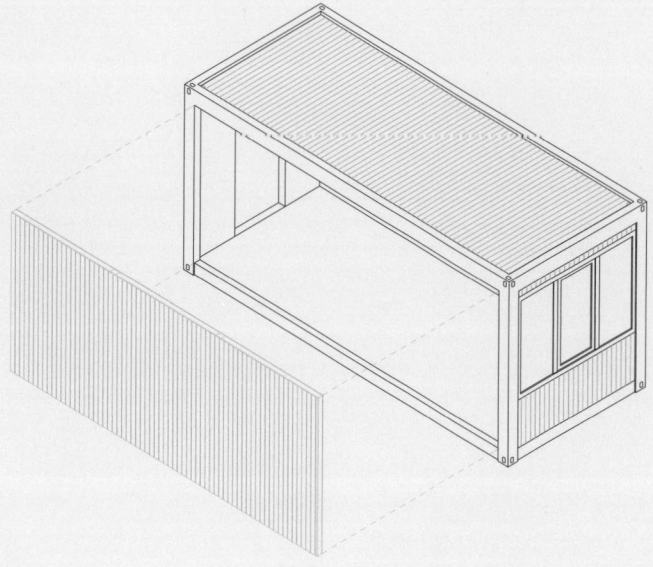

059 | Building container system

TYPE 2
BUILDING CONTAINERS

DEFINITION
The development of the building container involved the transferring of the concept of standardization and rationalization to a building module that would have the characteristic features of a freight container. The result is a prefabricated spatial cell module that is designed solely for building purposes and thus has a lighter structure.

STANDARD
Unlike a freight container, a building container already fulfills the civil engineering requirements demanded of a building envelope. However, it is a mass-produced item too, just like the freight container. In order to produce building containers as economically as possible, a standardized version in terms of fittings and structural design was developed that has established itself as a building module for short-term interim solutions.

The primary supporting structure of this standard version is made from steel profiles.

060 | Prefabricated steel components for container frames (photos from production at Kleusberg, from the company's Wissen site, Germany, hereafter referred to as Kleusberg)

The steel frame as a primary supporting structure is filled with a secondary supporting structure made of steel carrier beams that support the wall, floor, ceiling and roof fittings. The insulated surface fillings have a layered structure and an intermediate standard of thermal, moisture, fire and noise protection. They already contain openings such as windows and doors. Building containers are produced with a degree of prefabrication of up to 100%. They leave the factory ready to be put into service.

061 | Primary and secondary supporting structures (photo: Kleusberg, Germany)

Special versions of building containers use alternate materials in their supporting structures, such as aluminum, stainless steel, wood/glued laminated timber or fiber-glass-reinforced plastics.

The addition of a building container module leads to redundant double structures—e.g. with regard to the wall, floor and ceiling fittings—which are unnecessary from a structural design point of view. The advantage of standardized mass-production should be weighed up against the disadvantage of the space lost when combining modules. Fillings can be partially omitted in order to create spatial connections between individual elements. Openings in surface-forming components are only possible to a limited degree as the secondary support structure, which is necessary to ensure the stability of the frame, will then also be missing. Because of structural stability requirements, the basic module consisting of individual containers, always remains intact in a multi-story construction made of building containers; this basic module also determines the layout of the interior space. For this reason, larger joined-up spaces are only possible to a limited degree.

USE
If one considers the purposes of building container constructions, they are most often purely functional buildings that are used as an interim solution until a permanent extension is eventually built. Building containers can be used flexibly to the extent that they function as

Containers as building blocks

060

061

062

063

064

065

single rooms as all room-delimiting components are already present, but can also be combined to form multi-story buildings up to a maximum of 3–4 stories.

With their cell-like structure, these modules are suitable for layout systems with small rooms, which are then joined together. These buildings can be used for offices and administration, educational facilities (e.g. schools, kindergartens, and crèches) or medical facilities (e.g. laboratories, hospitals and mobile clinics, doctors' practices, intensive care units, ordinary wards).

Other possible uses for building containers include temporary accommodation (e.g. student accommodation, residences for asylum applicants, emergency accommodation in disaster areas), building site facilities (e.g. construction offices, storage and sanitary facilities) or technical equipment (e.g. equipment and air-conditioning containers, cooler containers).

062 | bbn – bed by night / Professor Han Slawik, Hanover, 2002

Building containers are also used as facilities for events, temporary sales buildings (e.g. ticket boxes, kiosks, fast-food restaurants), services (e.g. ATM machines), for information purposes (e.g. information stands, gatekeepers' buildings) and for presentation purposes (e.g. trade fair stands).

063 | seven-day architecture / LHVH Architekten, Munich, 2004
064 | derCube / LHVH Architekten, Freudenberg, Germany, 2007

Military applications are a particular area of use for building containers. There is a wide range of variants in terms of structures and fittings for these applications, which are tailored to meet the special technical requirements involved.[22] Special designs with higher security standards are also possible.[23]

Building containers are also used for research stations. Building containers can be transported in a fully-equipped condition if the research station is at a location with poor accessibility that can only be reached using special transports. Conventional construction work is often not possible at extreme locations because of the climate conditions there. The installation period available on-site may be so limited that the use of prefabricated modules is the only way of erecting a building during a certain time of year.[24] However, a second envelope must be provided around the building in the case of extreme climate conditions, as the thermal insulation of the building container will not be sufficient to buffer very high temperature fluctuations.

065 | Indian research station, IMS-Ingenieure and bof architekten, Antarctic, in progress

Containers as building blocks

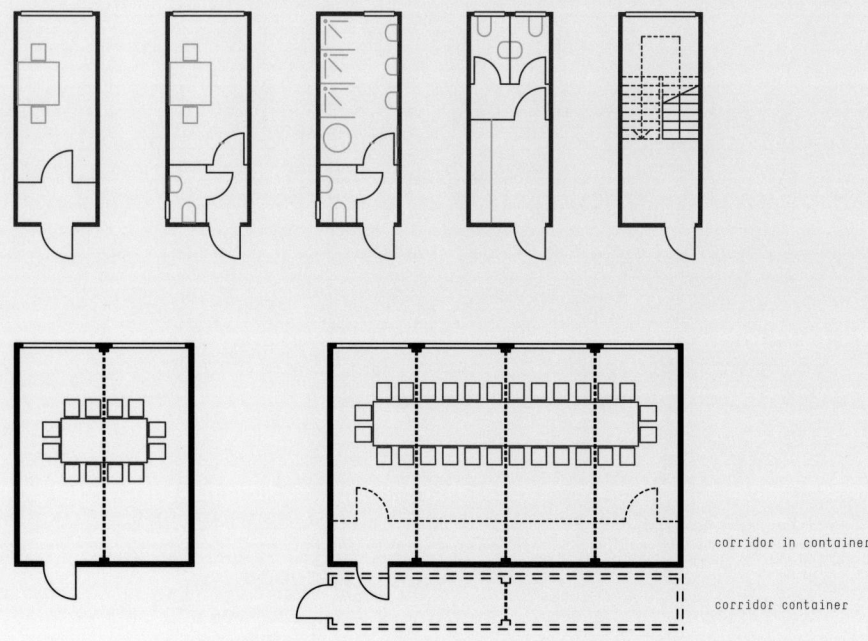

067

068

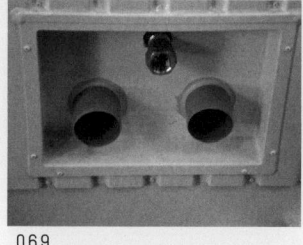

069

066

TYPES

Standardization that involves various types as universal solutions has become established in order to allow for industrial mass production. Building containers are thus available with various layouts and fittings. These are defined by manufacturers in terms of their uses and are even categorized into product lines.

The basic type is "one room" or "one room with additional room". These are standard office cells that are 20 or 40 feet long, and are either with or without an additional room (for filing or for sanitary facilities).[25] In the case of the "multi-cell room" basic type, a number of office cells are combined to create a large open-plan space or a number of individual offices. This combination can also be used for shower or bathroom facilities with a number of cells.

066 | Floor plan variants for one-room/multi-cell building containers

The stair container is used to provide access on multi-story buildings. It can be combined to create a multi-story stairwell.

067 | Production of a stair container (photo: Kleusberg, Germany)

Corridor areas can either be divided off using light walls or else created using corridor containers. Corridor modules may have differing dimensions (e.g. 1.5 × 2.5 meters or 1.5 × 5 meters).

068 | Corridor containers (photo: Alho, Germany)

A large number of special components are available that can be fitted to building container constructions. These include podiums, canopies, exterior ramps, corridor roofs, advertising boards on parapets, sun-protection lamellae and canopies to protect against sun and glare, fly-screens, additional sun-protection roofs, container roof platforms with rails and window flaps, and window security grills with clamp attachments. Roof reinforcement is necessary for more sophisticated roof fittings such as the planting of roof greenery.

The auxiliary technical equipment available includes air-conditioning and heating equipment (electrical indoor heating, gas or oil radiators—only when there is ventilation to the outside air), electrical installations, sanitary facilities, safes, equipment cabinets, sewage tanks and lifting equipment; these are either fitted to the outside (heating systems), slid under the containers (sewage tanks) or else positioned in a separate container.

Suspended ceilings with integrated lighting, floor reinforcements and raised floors for computer equipment can also be added.

069 | Utilities connections (photo: Kleusberg, Germany)

Some manufacturers offer building container systems with higher standards.[26] These systems have improved thermal, noise or fire protection, for example. Convenient room sizes, more generous facade design and interior fittings with higher-quality surfaces are all possible.

Containers as building blocks

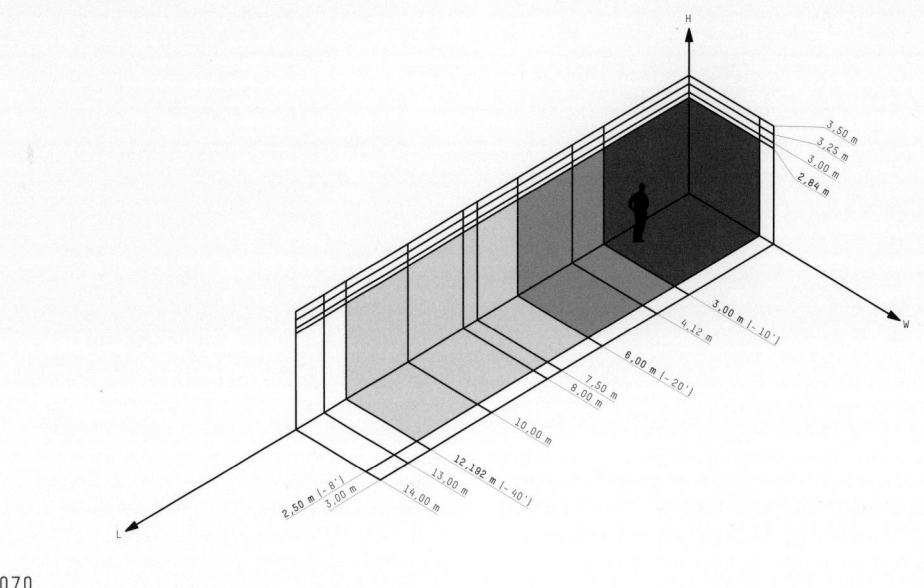

070

DIMENSIONS

Building containers are generally produced in line with the ISO dimensions that apply to transport containers. However, since manufacturers employ their own systems for production, transport and installation, it is also possible to use different dimensions that result in larger rooms, for example. These are based on the metric system instead of using dimensions in feet. In this way, the dimensions specified in the ISO standard have been extended. Building containers are not covered by an internationally binding standardization system.

The most frequently used standard version is the 20-foot building container. Other commonly used dimensions are 10 feet and 40 feet. Alongside the standard exterior width of 2.45 meters, there are also special widths of 3 meters and 4 meters. The height of building containers varies between 2.6 and 2.85 meters. For a height of 2.6 meters, the amount of head height available is only 2.3 meters. Larger special heights of 3 to 3.5 meters are also available to provide more head height.

070 | Dimension grid for building containers

071 | Table of dimensions for building containers

	Length ext.	Width ext.	Height ext.	Length int.	Width int.	Height int.	Usage area	Empty weight
10'	2,990 mm	2,438 mm	2,591 mm		2,232 mm	2,300 mm	6,5 m²	1,300 kg
20'	6,055 mm	2,438 mm	2,591 mm	5,852 mm	2,232 mm	2,300 mm	13,0 m²	1,950 kg
30'	9,125 mm	2,438 mm	2,591 mm	8,930 mm	2,232 mm	2,300 mm	19,5 m²	2,100 kg
40'	12,192 mm	2,438 mm	2,591 mm		2,232 mm	2,300 mm	26,0 m²	
Special width		3,000 mm						
		4,000 mm						
Special heights			2,800 mm			2,500 mm		
			3,050 mm			2,750 mm		
			3,500 mm			3,200 mm		
Special lengths on request	max. 12,000							

071

CONSTRUCTION

The primary supporting structure of building containers consists of a basic frame made of hollow steel profiles and rolled profiles; it is supported and reinforced by the secondary supporting structure as a substructure for the fillings. The welded steel construction consisting of the floor and roof frames is connected using four corner supports. The basic frame is pre-treated and either galvanized or painted to protect it against corrosion. Four corner fittings made of high-strength cast steel form the transport corners.

The fraction of steel in building containers is a lot less than in freight containers (the latter have around 70% steel). Fittings are responsible for a considerable portion of the self-weight of building containers. The relevant static loads that form the basis for sizing calculations for the supporting structure are the same loading scenarios that generally arise with tall buildings. These are the self-loads from the components, transport loads (depending on the type of use), and loads due to external influences such as wind, snow and ice. In addition, dynamic loading scenarios occur during transport and installation.

There are standard statics types for the sizing of building container constructions that already take all these loading scenarios into account. Project-specific statics calculations are only necessary in the case of special designs. Building containers can only be stacked up to a maximum of 3–4 stories because of the limited load-bearing capacity of the modules.

The supporting structure is then fitted with wall, floor, ceiling and roof fillings. The layered structure of these fillings varies depending on the level of comfort to be achieved. Once the frame has been produced, the fillings are created in the workshop largely by hand. The design of the fillings is precisely specified as part of the planning process. In this way, every container is tailored to fulfill a predefined usage. The amount of surface area occupied by window openings is relatively small for standard containers; these openings are implemented in the form of a perforated facade. Large-area, deep glazing is available in systems with a higher standard of fittings.

Galvanized crossbars that support a blind floor covering are included flush in the floor base as part of the floor structure. The floor is insulated with mineral wool and plywood sheeting is fitted on top of this, which then supports the homogeneous floor covering.

The roof consists of hot-dipped galvanized trapezoidal sheeting that is also supported by cross beams. These cross beams join up the frame structure to form a tortionally resistant unit. The roof takes the form of a trough roof, sealed using sheeting and then fitted with four roof inlets that are connected to rainwater drainpipes at all four corners. The roof is also insulated on the inside, fitted with a vapor barrier, and then covered with coated plywood sheeting.

The walls are finished using a profile sheet as an external sheet, which is then painted using a RAL color. The walls are insulated, fitted with vapor barriers, and then covered on the inside with coated plywood sheeting. Exterior doors made of steel are fitted to the door openings; the window elements are generally plastic windows with insulating glazing. The dividing walls are simple wooden-support constructions with insulation on the inside and cladding on either side.

072 | Wall-floor connection in a building container (photo: Alho, Germany)
073 | Ceiling-wall connection in a building container (photo: Alho, Germany)

PRODUCTION

Building containers are manufactured on a production line in 2–3 days in two main production steps: the first step involves the raw structure, while the second step deals with the fittings and finishing.

The steel parts are first cut down to size, derusted, and then welded to form a stable steel frame. The raw structure is complete once primer has been applied (production takes 2–3 hours per module).

074 | Welding the frame (photos: Kleusberg, Germany)
075 | Steel frame construction without fillings; rails and lifting equipment (photos: Kleusberg, Germany)

The fillings are then inserted into the frame. Horizontal surface elements (floor, ceiling) are first installed, followed by the vertical surface elements (walls). Window and door elements are also fitted as per the design planning.

076 | Surfaces being filled with thermal insulation (photos: Kleusberg, Germany)
077 | Installing windows and trapezoidal sheeting (photos: Kleusberg, Germany)
078 | Roof construction with drainage (photos: Kleusberg, Germany)
079 | Final coating of container modules (photos: Kleusberg, Germany)

Technical equipment is preinstalled or else fitted with sanitary utilities connections. Wiring is installed in each module and then connected for all modules later on. Interior cladding and roof seals are also fitted as part of the production process. The installation of fittings takes around 2–3 days. Rails and lifting equipment are used to move and rotate the container to aid the installation process. The container is ready for service once painted with a two-component RAL[27] acrylic lacquer or a plastic coating.

TRANSPORTATION

Building containers are fitted with devices necessary for transporting modules and fastening and connecting them to one another. Manufacturers generally transport their building containers on manufacturer-specific platforms whose dimensions are tailored to suit those of the modules. The building containers are fastened to the platforms using belts and special locks. They are then transported to the site and installed by installation technicians. Companies use mobile hydraulic cranes and installation transporters here. Building containers have crane lugs at the corner attachments that serve as lifting aids when installing containers using a crane. Two 20-foot containers can generally be transported on each truck trailer. Pack systems that can be taken apart represent an exception here, as they allow for the simultaneous transport of a number of building kits on a single truck.[28]

080 | Crane lug for crane installation (photo: Alho, Germany)

Building containers have corner reinforcements for the transportation of modules and the combination of spatial cells by stacking them; these reinforcements are comparable to the corner elements on freight containers.

Manufacturer-specific connection devices are available to securely attach spatial cells; they are similar to the rougher, more robust devices used on freight containers, but are much lighter.

081 | Vertical connection device (photo: Alho, Germany)

Building container modules are also linked together by means of utilities connections. Installation strands are combined over all the modules and then connected to the supply station.

082 | Utilities connection/electrical cable strand (photo: Alho, Germany)

072

073

074

Containers as building blocks

075

076

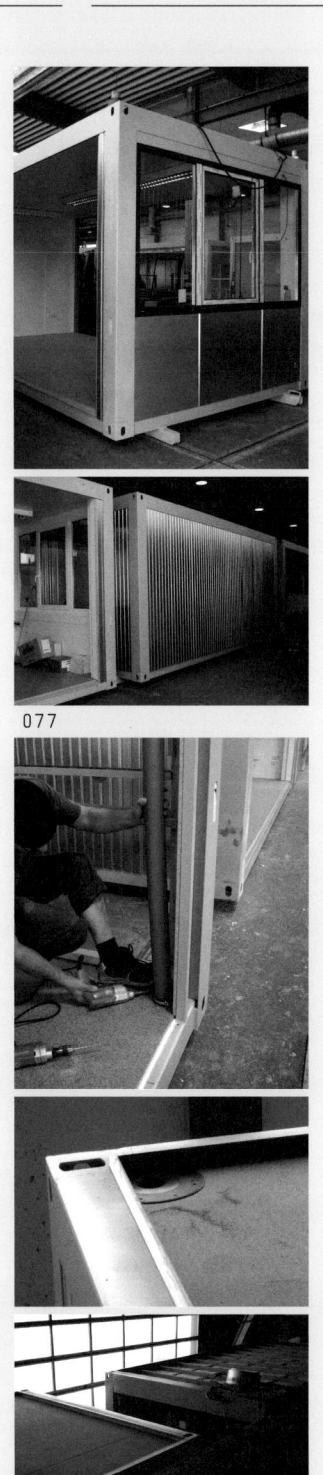

077

078

079

080

081

082

- 35 -

VARIANTS

In order to minimize loads, light containers do not use a steel frame structure, but instead have stable wall, floor and ceiling elements. They are particularly suitable for installation on sites with poor accessibility. They also have a substructure that can be leveled. They can withstand being erected and dismounted at least 30 times, and can be installed by two people in 90 minutes without lifting gear.

Field camp containers are built in a particularly robust and stable manner, and their basic structure and walls are made of steel. The fillings are welded to the frame so as to create a seal. The metal components have long-term corrosion protection. As their name suggests, these containers are manufactured for use in field camps.

Folding containers have light structures and also save on space as they can be dismantled for transportation. They are also made of steel, just like standard building containers. When folded together, they are around 70 centimeters high. In line with their special function, they are suitable for quick, flexible use for applications where space savings during transportation are an important consideration.

The principle of the folding container was used in the "flexotel" in the Netherlands. The concept involves hotel-room units that can be erected at any given location desired. They are generally used for temporary guest accommodation for events. Structurally, this container is similar to a building container, but has the special feature that it can be folded together. It is fitted with freight container corners rather than with the corners typically found on building containers. Two units make up a single hotel cell. When folded together, the containers can be efficiently transported as a number of units can be packed onto a single truck. These are then "folded out" again on site using a crane.

083 | Folding containers/foldable hotel unit

Extendable containers (3-in-1 system) are the same size as a standard container when folded up. These building containers can then be extended on site using a folding mechanism. Installation only takes a few minutes. These extendable units include a 2-in-1 version and a 3-in-1 version, up to a maximum of a 4-in-1 version.

084 | 3-in-1 version of the extendable container

CONSTRUCTION PHYSICS

There are inherent structural weaknesses in the building container system. The high thermal conductivity of steel means that there are thermal bridges where the insulation is interrupted or reduced by the structure. This occurs mainly at the corners of a building container. In the case of permanent usage, this can lead to reduced comfort and to damage caused by moisture. After a longer period of service, leaks in the roof can also be expected.

These structural disadvantages are tolerated in the case of building containers so that the detailed solution used can be kept as simple as possible. Building containers are generally more suitable for temporary usages because of these weaknesses. An insulated building container still represents a significantly better civil engineering solution than a converted transport container.

Alongside regulations regarding thermal protection, there are also other guidelines that specify fittings standards, particularly in the case of workplaces. A minimum standard of fittings is necessary in order to fulfill these requirements. [29]

Production is subject to predefined quality standards. The warranty period for building containers is not uniformly regulated and it can vary depending on the manufacturer (1–2 years).

ECONOMIC ASPECTS

Certain manufacturers provide production, development, rental, sales and service for finished containers as a complete package: they carry out all tasks including consulting, planning, permission from the authorities, as well as construction itself and management of construction. The end product is then a turnkey facility that includes building services equipment, installation and interior fittings, which can be either bought or rented. Depending on the size of the order, the production period is around 4–6 weeks. Furnishings can also be provided as an additional convenience for clients. Manufacturers' websites feature detailed information with price-calculation tools, online planning systems and customer hotlines. Alongside rental and purchasing, manufacturers also offer financing, rent-to-buy and leasing.

Building containers are maintained and repaired in the workshop every time they return from rental use before being rented out again. Certain companies also have depots that can be used to store containers. Rental containers are subjected to regular checks. Every container returned is first given an "in-check" before being cleaned and stored in the depot. Containers are cleaned on cleaning lines in the workshop. A logged "out-check" is carried out when the container leaves the depot in order to document its condition.

The cost of a structure made of building containers is around 30–50% of that for a conventionally built building, on average. The costs are lower for containers placed in larger halls as they do not require climate shells.[30] The issue

083

085

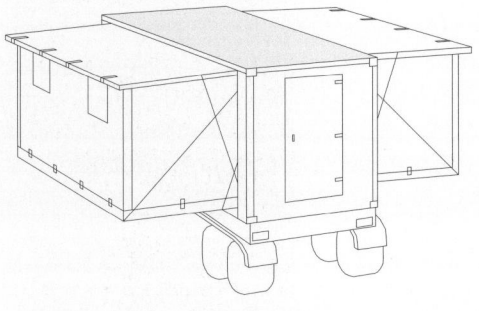

084

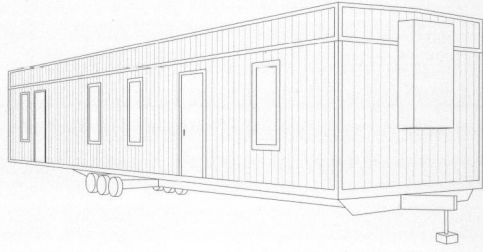

086

of whether it is better to buy or rent a container depends on the requirements of the client and on the planned service life. Some manufacturers prescribe a minimum rental period of between three months and one year. The average service life of a building container is around 15 years[31], and even longer lives can be achieved for light structures at permanent locations. Once the service life has expired, the building containers are sold to countries where used containers are still in demand.[32] Containers no longer fit for use are recycled to recover their materials of construction.

The cost of a building container depends mainly on the fittings standard used. Very simple rental containers cost around €10 per square meter per month, while more sophisticated variants cost around €30 per square meter per month, or more. Installation costs for erecting and disassembling containers are around €15–25 per square meter. At the end of the rental period, cleaning costs also arise depending on the amount of soiling. The total rent often exceeds the purchasing price for rental periods longer than around 36 months, meaning that purchasing is the more economic option. However, this should be checked on a case-by-case basis.

New building containers cost between €550–750 per square meter, meaning that a 20-foot container costs between €5,000 and €10,000. Used versions of building containers cost around 50% of the price of new containers. However, it should be considered when buying containers that the purchaser will also have to bear disposal costs later on.

SPECIAL FORMS

Standard building containers have particular disadvantages with regard to architectural design and room quality. For this reason, the manufacturers of standard modules also include systems with a higher standard in their product ranges, although these have not proved competitive on the market so far. Room modules similar to building containers with alternate materials, designs and dimensions can also be found on the market. They can all be broadly classified as building containers as they are modular, spatial-cell systems that can be transported in the same manner as freight containers.

Generally, these special solutions are tailor-produced spatial modules that have been developed for a particular project. The potential of these individual solutions is very promising with regard to the future development of building container systems as a high-quality architectural solution.

085 | HomeBox / Professor Han Slawik, Hanover, 2008

A comparison of the situation worldwide reveals that the types of building container on the market vary from country to country. For example, office and living containers are widespread in the USA—these are so-called trailers that already have their own chassis and function as mobile units. Their wheels are generally removed once they are in place at their final location and the trailers are then mounted on fixed foundations. The gap between the ground and the trailer is blocked off by an apron.

These trailers are used in the same way as building containers in Europe, and are made of wood and cladded using aluminum sheeting. Unlike building containers however, they cannot be stacked, and can only be combined horizontally. Thus, they do not count as a building system.

The use of trailers as both offices and homes is not unusual in the USA. Some Americans live in "trailer homes" that are either mounted individually or else as part of temporary estates. The lack of insulation, poor-quality and leaky windows give trailers a negative reputation of being inferior homes.

086 | Trailer, USA

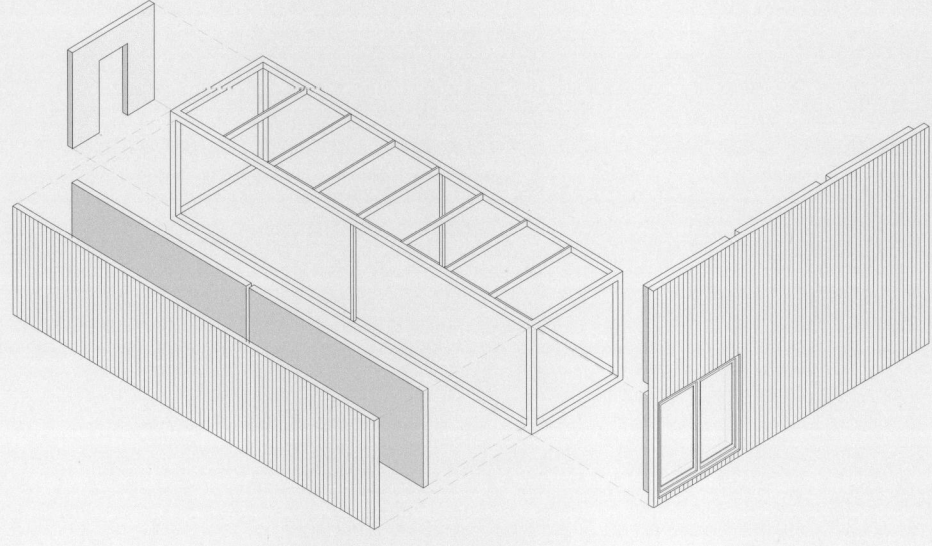

087

087 | Modular frame construction method

TYPE 3
CONTAINER FRAMES

TYPE 3A:
Modular frame systems

DEFINITION
In principle, the modular construction system with frame cells is similar to the building container system described previously. The supporting structure of the modules is formed by a stable frame, just as in the case of building containers. It is industrially prefabricated in the factory as a compact, mobile unit along with part of the fittings, and is then assembled at the construction site to yield an overall building. However, in contrast with building containers, the building's "finish", which may consist of a weather-resistant building envelope or floor and wall coverings, is created and installed on site for all modules together. This means that a building made from module frames may take on certain design features of a conventionally constructed building. However, combination with conventionally produced components does reduce the ability of the building to be disassembled and reassembled.

Modular frame construction is significantly more flexible with regard to space, floor plan, and facade than previous container construction systems. Module frames are tailor-produced for the project in question, in line with the specifications ordered. Accordingly, the dimensions can be freely chosen here, although manufacturers also offer standard sizes that have proven to be suitable for certain usages. Although transport considerations do limit the maximum size of the modules, these dimensions can nonetheless still be quite generous as transport is carried out using custom-made platforms. It is not planned to integrate modular frame systems into the ISO-specified transport system, which reduces the flexibility of transporting modular frames. The dimensions and the fittings of the individual spatial modules are both implemented according to the user's wishes with the required degree of prefabrication.

DEVELOPMENT
The first modular frame systems were developed as early as the 1960s. The impetus for the use of modular frame systems was provided by short-term increases in space requirements. Spatial-cell systems were mainly to be found in Europe and Japan. A Japanese company called Sekisui developed a fully independent construction system called "Heim". This was a system for building homes. Some German manufacturers of building containers, such as Ofra (temporary classrooms), Alho and Kleusberg, were also involved in these developments. They drew on their experience and expertise from container manufacturing and were able to use their existing production facilities and assembly lines in cases where modular frames were developed based on container frames.

Building using individual steel components that are delivered separately and assembled on site is much more widespread in the USA than in Europe. Systems based on frame cells are less common in the USA and have only begun to appear on the market in recent years. In Japan, however, spatial cell systems made of steel have been more successful than frame systems in the modular building sector. In Germany and the rest of Europe, the spatial-cell construction method is widespread, and wooden frames are becoming increasingly popular too as a cost-effective, long-term solution.

USE
Building using modules is most suited to repetitive building complex projects. For example, hotels are often built as frame

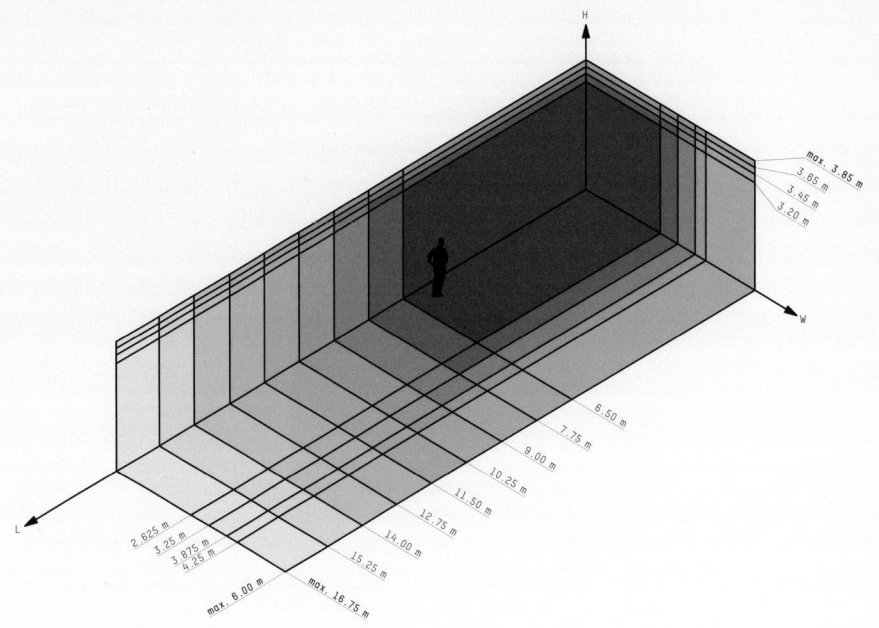

088

modules as they have a large number of identical rooms. Buildings based on prefabricated solutions also represent a cost-effective alternative to conventional buildings for schools, offices and administration buildings, laboratories, hospitals and multi-story residential developments.

The modular construction method allows for extensions and conversions later on. The supporting frame structure also represents a relatively flexible, adaptable spatial skeleton. For example, inside walls can be moved to a certain degree, even after a long period of usage, in order to alter the rooms to meet changing requirements.

Although individual room modules can be removed and reused, this does involve quite a lot of work due to the conventional components that have been added. For this reason, modular construction systems are mainly suited for permanent buildings.

DIMENSIONS

It is difficult to directly apply container dimensions to universally applicable building modules, mainly because of their insufficient width of around 2.44 meters. A special width of 3 meters has now become established for building containers, as this makes them easier to furnish, even if it does also make transport more difficult. However, the dimensions of frames are determined by logistics and by production conditions. On the one hand, manufacturers' existing production facilities lead to limits on sizes connected with continuous production on assembly line trolleys; on the other hand, the dimensions are also limited by the transport vehicles used and by road widths. Dimensions of 20 meters for the length and 6 meters for the width are actually theoretically possible for frame structures made of steel[35], if one were only to consider static constraints.[34]

Modular buildings generally do not have to fit in with a predefined building grid. Frames are tailor-made for the project in question and are available in widths that can be freely chosen between 2.5 meters and 4.5 meters, with room heights up to 3.5 meters and lengths of up to 18 meters.[35] System manufacturers also supply certain predefined module types. Since they can be combined horizontally and vertically, it is a good idea to limit the selection to standard types, meaning that modules are combined as part of a system. Within this system, the flexible applicability of the basic modules makes it possible to realize various floor plans and building forms.

088 | Dimension grid for module frames

PRODUCTION

The production principle is similar to that of building containers. The load-bearing frame structures are manufactured on assembly lines and equipped with largely complete interior fittings, including technical utility lines.

089 | Prefabricated frame cells (photo: Alho, Germany)

In this way, frame cells with interior fittings can be produced with degrees of fabrication up to 90%[36]; however, the degree of prefabrication is only around 60% for the overall building on average. In contrast with the usual situation for building containers, components that are exposed to the elements, such as facade and roof, are installed on site for all modules together. This is also possible in the case of wall and floor coverings. This results in a combination of prefabrication and conventional construction methods. Some manufacturers also provide prefabricated and partially prefitted elements for the facades.

The high degree of industrial prefabrication and the ability to work independently of the weather in the factory can significantly reduce the construction period. However, the actual amount of work involved is determined by the quality and type of design desired by the user.

CONSTRUCTION

The main supporting structure is formed by the supporting frame with eight beams and four corner supports (superstructure). Depending on the requirements for fittings later on, the frame is fitted with a supporting substructure in the divider area. Floor and roof elements are usually made from profiles with varying thicknesses. Depending on the module length, it may also be necessary to use intermediate supports. Even though the main loads are supported by the frame, the additional secondary substructure is still essential. Steel has become established as the material of construction of the load-bearing structural elements. Systems with wooden or steel-bond structures are also available, however.

The modules are standardized, statically identical units that can be ideally adapted to their subsequent function and their role within the overall building. Clearances, stairwells and large, support-free room widths must all be taken into account during the manufacturing process. The individual room modules can be combined at their ends or longitudinal sides to create a complex supporting structure and, in contrast with building containers, they can be stacked to form up to six stories.

Although similar in terms of type and design, module frames are not as flexible as the system made from container frames that is described in the following. Adapting the frames as much as possible for the subsequent use makes it difficult to alter the existing building later on, as the additional supporting structures in the open areas mean that the fillings can only be removed to a limited extent.

090 | Prefabricated frame cells (photo: Alho, Germany)

As part of the fittings to be carried out, fillings for walls, floors and ceilings are inserted into the open areas in the supporting frame structure. Light steel profiles with mineral fiber insulation, an inner vapor barrier and gypsum board cladding have become established as standard fittings for filling the external walls. Inner walls and ceilings also use drywall construction with supporting structures and planking. The standard floor construction consists of galvanized profile sheets with insulation and a floating dry screed on top of a supporting board. All building elements such as windows, doors, interior doors, stairs and all building services equipment are delivered to the building site with pre-applied surface finish and are then mounted.

The facade is usually mounted on site too. This involves a continuous insulation layer in accordance with the relevant building physics requirements, with plaster or facade cladding, and with back-ventilation if necessary. The continuous facade, which is common to all modules, covers the entire frame structure and thus minimizes the thermal bridges.

The roof construction with sealing is realized with a liner sheet that is fitted as a sub-roof and a protective roof that can be freely selected and is fitted on site. With the modular construction method, a range of roof forms such as flat roofs with slope insulation, monopitch roofs, gable roofs or arched roofs in various materials are possible.

TRANSPORTATION

Construction begins by creating the foundation at the installation site. Flat-bed trucks transport the modules to the building site, where they then need to be mounted and connected to the electricity, water and sewage grids.

The modules are generally mounted after a planning and production period of around eight weeks. Although installation of the spatial modules can be completed within 2–3 days, it still takes a number of weeks for all fittings, finishing and for mounting of the facade. The units are thus generally ready to be used 10–12 weeks after being ordered.[37] However, the actual construction periods depend on the scale and design of the project in question.

Unlike other container building systems, module frames generally do not have transport openings at their corners. They are transported on the manufacturers' own vehicles, and are secured using tension belts. These are often oversized transports because of the sizes of the frames involved. However, ease for transportation is only of secondary importance for this system. Once transported to the building site, the modules are fixed in place, and are then joined up using conventional means, without the use of special attachment devices. This reduces their mobility and means that they can only be considered as a container-like system.

091 | Transportation and mounting of spatial modules (photo: Alho, Germany)

CONSTRUCTION PHYSICS

The manufacturers' intention when developing the modular frame construction method was to develop a modular building system with optimal construction physics characteristics. Only in this way can unlimited applicability for these buildings be guaranteed without completely losing the advantages offered by prefabrication.

090

091

This adaptation to higher requirements was necessary in order to establish modular construction for applications outside of temporary usages. The increase in insulation thicknesses and the use of materials with better insulating properties were mainly responsible for the fulfillment of these requirements. The thermal bridges in the frame areas were avoided by pre-mounting the building envelope.

In this way, buildings with modular frame construction systems can adhere to all requirements resulting from standards and construction law that are demanded of permanently erected buildings with regard to construction physics and noise and fire protection. The modular frame construction method also makes it possible to attain the standards achieved by conventional buildings.

ECONOMIC ASPECTS

The costs for a building erected using modular frame methods start at around €600–700 per square meter and can go as high €1,500–1,700 per square meter. These costs apply from the top edge of the foundation upwards; foundation costs should be considered separately. The construction costs are also strongly dependent on the client's specific wishes and on the finishing standard required for the building. In principle, this construction method results in lower costs mainly due to the rationalization of the building process and the associated shorting of the production process; the costs for the structure itself are approximately equal to those for a building of a similar standard built by conventional means. The modular frame construction system benefits from the advantages of modular prefabrication. Nonetheless, the disadvantage of high material use due to the redundant doubling of certain components still exists, and money cannot be recovered by reusing components either.

ECOLOGICAL ASPECTS

In terms of ecological efficiency, modular frames are not particularly favorable. Re-usability is very limited, as certain components are added to the mobile, prefabricated spatial cells in a conventional manner and would have to be produced again in the case of reassembly of the structure at a new location. The removed building elements, such as the facade, have to be disposed of.

The flexibility and variability of a construction system should also be considered according to environmental considerations. In this system, the secondary supporting structure is essential, meaning that the modules are not as flexible as a completely free system with non-load-bearing fillings alone, such as the system with container frames. In general, it is not possible alter the building with modular systems (except for moving inner, non-load-bearing walls).

Containers as building blocks

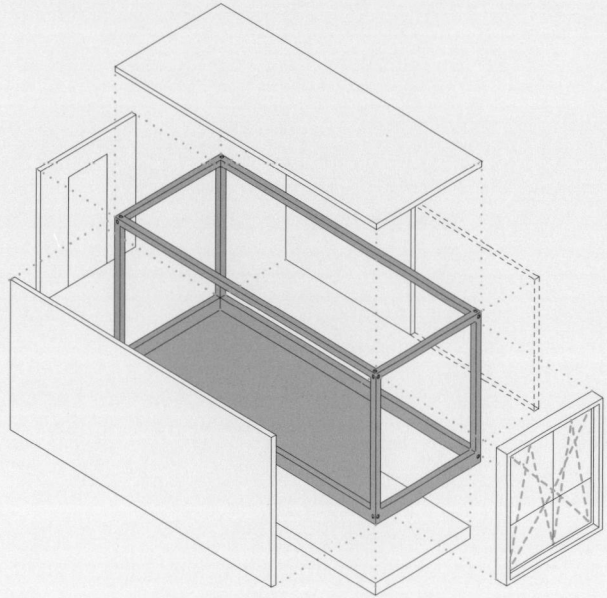

092 | Container frame system

TYPE 3B: Container frame systems

DEFINITION

Against the background of the container building systems described so far, the idea arose to create modular building system that was as flexible as possible in terms of spatial, structural and design considerations, with a high degree of prefabrication and which, like the systems discussed so far, would also be based on the container principle.

There are already similar systems on the market from various manufacturers, but they generally only have limited flexibility and variability: the fillings can only be moved laboriously and with statics limitations, as they always perform a load-bearing function too. Certain components become redundant when the modules are combined (double walls, ceilings and floors), which increases the amount of materials needed and results in wasted space.

In order to optimize the container construction approach, the author of this book, Professor Han Slawik, developed his own building system using container frames. The result was the idea of separating the load-bearing frame from non-load-bearing fillings, i.e. the separation of supporting structure and fittings.

This system with container frames is currently still in its development phase and is mainly intended as a demonstration of the potential of modular building systems based on container frames. It is planned to build prototypes together with cooperation partners that can be used in smaller units and also combined to form larger building structures. This self-developed system has already been used and refined as part of various studies and competitions. This building system has already been patented.

DIMENSIONS

Two particular features of this building system are that it adheres to the preferred dimensions specified in the international ISO standard, and that the transport corners are fitted with devices that allow both for transportation using standardized transport equipment, and for the modules to be easily connected to each other. Additional attachable modules with the same dimensions as standard ISO modules are possible for various uses, such as access zones, service units (sanitary rooms, small kitchens), storage rooms, conservatories, etc.

USE

Various building structures can be achieved using the supporting module. The room-delimiting elements can be used to create rooms of varying sizes and heights. Different room layouts can be implemented depending on the usage requirements—e.g. as an office, studio, atelier, workshop, laboratory or temporary accommodation. This ensures variability in the fittings and flexibility with regard to uses, i.e. the continuously changing needs and habits of the users are taken into account and it is possible to alter the rooms to suit changing circumstances.

CONSTRUCTION

A supporting frame made of steel tubing (square or rectangular) forms the supporting structure; however, depending on requirements, this frame can also be made of aluminum or wood (glued-laminated timber), or of prestressed concrete rods or composite materials. The container frame supports the entire load of the building and is therefore more generously sized

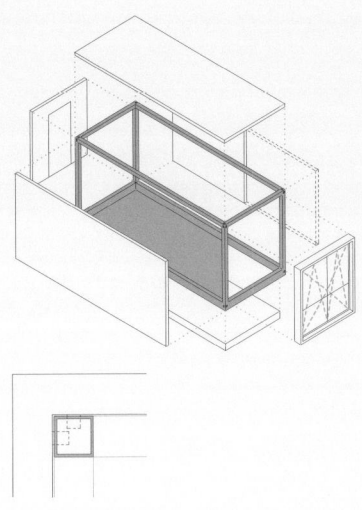

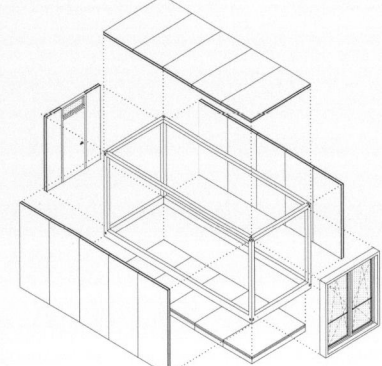
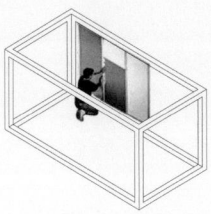

093

095

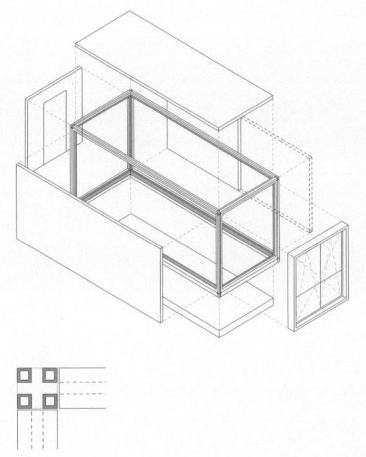

094

than the frame of a building container or a module frame, for example.

093 | Facade in front of a one-piece frame (variant 1)

094 | Facade mounted flush in a thermally separated frame (variant 2)

The frame filling is available in two variants: with the first variant, the facade is fitted in front of the frame, meaning that the frame is covered over by the facade elements. This construction avoids thermal bridges, and the outer dimensions of the filled frame are increased by the insulation thickness of the facade elements. Under certain circumstances, transportation can be restricted when a premounted facade is present. The profiles of the supports and beams can be one-piece items.

The facade is installed flush with the surface in the second variant. The modules remain fully transportable and can be moved more easily later on. However, thermal separation of the container frame components is necessary in order to have the wall fillings at frame level. To achieve this, the supports and beams must be made of a number of smaller parts. There are various ways of achieving thermal separation. Concept studies on this subject will be continued together with cooperation partners as part of the next project phase.

In supporting frames, the frames are not load-bearing and act as simple room-delimiters. This means that freely selectable spatial connections can be created in the horizontal and vertical direction within the frame structure. It remains possible to remove and remount the frames and fillings, and the fillings can be altered independently of the supporting structure. This means that structural alterations that do not involve excessive labor or costs can be carried out during the construction phase or else later on. The doubling-up of surface-forming components (such as walls, ceilings etc.) does not occur with this system. As with the previous systems, production can be carried out in a workshop in order to minimize the construction time on site.

095 | Finishing the frame modules using prefabricated elements (walls, floors, ceilings)

The fittings in the form of non-load-bearing, room-delimiting components (floors, ceilings, outer walls, inner walls) can be prefabricated according to standardized dimensions and then mounted, disassembled and remounted using a simple click system. Smaller elements and module-scale boards (e.g. wooden-frame construction elements) are all possible.

These elements can be selected as standard finished components from a catalog; however, the possibility of designing individual frame fillings should also be retained, e.g. by selecting specific materials and surfaces that are available from builders' providers. The simplicity of installation means that users will ideally get involved in the planning and building process themselves and introduce their own specific building materials for non-load-bearing room dividers (wall, ceiling, and floor). In this way, the building users will be able to make the building "their own".

As the container frames are combined with commercially available building elements, this is system can be considered to be an open building system.

096 | Combinations with additional modules

Containers as building blocks

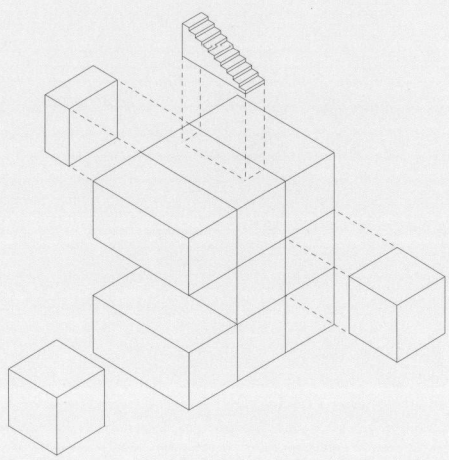

The flexibility of the building system allows for a wide range of combinations, including additional modules, and creates the basis for design freedom involving various materials, colors and surfaces. Horizontal and vertical interconnections and open spaces turn the mere stacking and arrangement involved in container construction into real architecture, with the help of these room-defining measures.

TECHNICAL FITTINGS

Installation connections are generally integrated into the walls in office containers, making it difficult to alter or replace them later on. For this reason, we recommend standardized and compact, rod-shaped elements with installation lines and plug connections to supply upper installation areas. Lower areas can be supplied using cable tracks, and it should also be possible to implement all circuits in a wireless manner.

If no supply and waste removal infrastructure is present on site, technical modules (power packs) on a modular scale (basic modules and/or auxiliary modules) can help provide self-sufficiency. It should also be investigated whether rainwater can be collected and used as gray water.

ECONOMIC ASPECTS

The investment costs for a building system that offers maximum flexibility and variability will certainly be higher than those for the previous building systems. However, as this system is still being developed and is not yet available on the market, it is not possible to quote definitive data regarding costs. The more generously sized supporting steel frame has an influence on the costs. In particular, the costs for the solution with thermally separated frames cannot yet be quantified. For fillings, only the increased labor costs for the reversible click system have to be considered as conventional building materials (such as sandwich elements) are used here in all other cases.

The investment costs must also be considered from a sustainability point of view. In this way, higher initial costs could pay for themselves over a longer period as significant investments for conversions, extensions, and even for reuse or new construction will not arise or will be more modest due to the ease with which the building can be modified.

ECOLOGICAL ASPECTS

From an environmental point of view, this is a very sustainable building system with great development potential. As with a modular system based on building components, all components are joined in such a way that they can be separated again, can be removed with minimal effort, used in different ways, or recycled. It is possible to react quickly to changes in the requirements demanded of the building or to changed usage circumstances: the size of the building system can be extended or reduced, if necessary. The inner walls are all non-load-bearing, and the ceilings too, only support their own loads and are ideally just as moveable as window elements and facade parts are. The flexibility of facade design makes it possible to achieve a varied appearance and to use materials that are favorable from an ecological point of view. The system allows for the facade to be renewed if damaged or if thermal insulation requirements become more stringent.

The service life of the building elements should be carefully considered: under certain circumstances, the steel frame may still be intact after the fittings elements have been removed and it could be reused as a basic frame. The potential of this building system is quite considerable from an economic and environmental perspective.

REMARKS

1. For example, a maximum width of between 2.5 and 3 meters applies for road transportation, depending on the type of road and the nationally applicable regulations. Objects that are wider than 3 meters count as special loads and must be accompanied by a suitably marked private vehicle in Germany and other countries.

2. The word "temporary" means "For a limited time, ephemeral, not constant; transient", source: http://en.wiktionary.org/wiki/temporary, as of: November 29, 2009.

3. Source: Lower Saxony building regulations 2003, "Genehmigungsfreie bauliche Anlagen und Teile baulicher Anlagen", Appendix 11: "Fliegende Bauten und sonstige vorübergehend aufgestellte oder genutzte bauliche Anlagen".

4. Transport by ship €0.07/km covered (Distance Hong Kong–Hamburg is 9,026 km, approx. €600 incl. customs and port fees).

5. Cost of road transport by truck €2/km covered; special transport is more expensive.

6. Refer to the applicable depreciation write-off in Germany according to the depreciation tables at www.bundesfinanzministerium.de.

7. The purchase price for scrap steel was €200-250/t in 2008; in 2009 it is only around €60/t because of the fall in the price of steel.

8. Kienzle.

9. Levinson, Marc; *The Box*; Princeton 2006.

10. As per ISO 668, from: *Freight containers-ISO standard handbook*; Geneva 2000.

11. DB AG (German State Railways).

12. TIS, the Transport Information Service of German transport insurers.

13. The container types are defined in the ISO standard and described in the following sources: *Freight containers – ISO standard handbook*; Geneva 2000 and Hapag Lloyd AG.

14. Survey of prices from various suppliers in March 2009 (Renz Handels- und Logistik GmbH, Finsterwalder Container GmbH, Containex GmbH, Conical GmbH).

15. Sietz, Henning; "Emma Maersk – Heimat für elftausend Container", in the *Frankfurter Allgemeine Zeitung*, November 14, 2006. Estimates of the ship's capacity vary between 11,000 and 15,000 TEU.

16. www.hafen-hamburg.de.

17. "USA rechnen mit Einbruch im Containerumschlag" in: www.verkehrsrundschau.de, February 17, 2009.

18. Press release from DB AG (German State Railways) from January 24, 2008.

19. Rautaruukki Oyj, Rakennustieto Oy; *COR-TEN Facades*; Helsinki 2001.

20. Nikolaus, Katrin: "Wächter mit tausend Augen", in: *Pictures of the Future*, 2007, Siemens.

21. www.sec-bremen.de.

22. An example of a manufacturer in Germany is www.drehtainer.de.

23. Containers with bullet-proof designs (resistance class FB3) are also available to meet special safety requirements.

24. For example: Indian research station (bof architekten) or new Neumayer III building.

25. Special types, such as WC containers, are supplied in lengths of 5 feet or 8 feet.

26. For example: Alho basic plus, www.alho.com.

27. Special finishes in RAL colors are possible.

28. e.g. CS Transpack system, source: www.csraum.de.

29. In Germany: Workplace Ordinance (ArbStättV).

30. Hall inserts: Purchasing price approx. €220 per square meter.

31. In comparison: 50 years for a permanent building erected by conventional methods.

32. e.g. former Eastern-Block countries.

33. Stahl-Informations-Zentrum, Brochure 573, "Stahl im Wohnungsbau – innovativ und wirtschaftlich", Düsseldorf 2002, p.23.

34. Stahl-Informations-Zentrum, Brochure 548, "Kostengünstiger Wohnungsbau mit Stahl", Symposium, Dusseldorf, September 22, 1998.

35. According to: standard dimensions on www.kleusberg.de. The dimensions depend on the manufacturers in question. Wooden frame modules are only available at lengths of up to 14 meters because of their load-bearing capacity (www.induo.de).

36. Almost all module manufacturers advertise a degree of prefabrication of 90%; however, this always refers to the prefabrication of the modules with regard to interior fittings.

37. Construction period for an administration building with around 1,000 m² of usable floor area (www.alho.de).

06

STRUCTURAL DESIGN ASPECTS
GUEST CONTRIBUTION FROM THE ENGINEER DOUWE DE JONG

STRUCTURAL ENGINEERING AND FREIGHT CONTAINERS

In this part an approach is given as to how to calculate structures assembled with containers. Also covered is how to handle these units themselves and how to change or adjust the structure of these units according to their new purpose.

Building with units in general

Buildings assembled with freight containers must meet building codes and rules which are valid in the country where the building is going to be built. The building must also fulfill the requirements of the owner and the user. These requirements can be different for each project and will not be part of this chapter. We will not discuss foundations and other general parts of buildings outside the assembled units as staircases because in general they are the same as for conventional buildings.

Freight containers

Freight containers are self-supporting structural units. They are available in specified volumes and in many designs. Most of them are used as freight boxes. A worldwide set of rules and demands has been developed for these containers concerning handling and shipping. Because of this standardization, shipping costs are lower when containers/freight boxes meet these specified requirements. These requirements are specified in International Standards. The only really standardized parts of these ISO standardized freight containers are the corner fittings, i.e. their exact place and the way they are placed. These corner fittings are mostly on the eight outside corners of the container, four on each side, and that is where the container meets the outside world. These standards are based on transportation and handling equipment. The other elements between corner fittings are up to the designer, and to the requirements of the client. Mostly these consist of a floor, two walls, a roof and a front and rear side, all within the space in-between the eight corner fittings.

ISO test loads and structural calculations

The requirements for containers are documented in so-called ISO (International Standard Organization) codes. An ISO handbook is available for freight containers in which a number of ISO codes concerning freight containers are published. The requirements that each freight container must meet in order to be shipped in standard ways, and for the lowest possible shipping costs, are specified in these codes. When freight containers meet these ISO codes, they are marked as certified and one can expect that they perform up to the required loads, and within the required deformations. These ISO codes are valid in all ISO member countries, and authorities in those countries must accept the capacities for loads and deformations as specified in the ISO codes. The manufacturers of freight containers manufacture and deliver their containers according to these ISO codes.

By doing so, and by tests according to ISO codes, their product is certified. This is done by test loads conforming to the ISO requirements by accredited companies like Lloyds or other qualified suppliers. Structural calculations will be less specific than test loads, so a great part of approving is carried out through test loads rather than structural calculations. And calculations give less result or ask for more material then the result of testing does.

When nothing is changed structurally in a certified freight container, it can be a building unit in itself. This gives every user or structural engineer specified information about the unit capacities for vertical and horizontal loads and deformation tolerance. Deformation is maximized for freight purposes and handling and storage tolerances for stacking. They can be calculated into elastic deformations and this information can be used to demonstrate building deformation requirements per unit. When stacking more units in heights or in a row, these units work together when structurally combined. Calculations of strength and/or stability according to the valid standards based on the steel specifications can also be made to demonstrate that a container building meets the requirements. But the outcome of structural calculations will most likely be less than the results of test loads. In a lot of cases, when new types of containers are developed or designed and the numbers are sufficient, test-load proof of required capacity will be used instead of structural calculation.

Assembling units

Building with containers means to assemble a number of units into a building or a structure. Assembling units means that connections are needed and these connections must be designed according the local building codes and the requirements of the client/user. The strength of these connections between containers in a building must be calculated, demonstrated, and designed according to local building codes and according to what is necessary for the actual building to be economic and safe.

Structurally assembled units

As with any building, vertical weight loads, and horizontal loads from wind, must be transported to the foundation, and/or outside bearings, and/or stability elements, by the designed structural elements. These are basic calculations and are the responsibility of structural engineers and designers of any kind of building all over the world. For standardized connections between containers and corner fittings, a number of shipping and storage parts are available and used by container companies during shipping, storing and handling. The capacities for loads and forces of these connecting parts are defined by their suppliers or manufacturers and are standardized and commercially available worldwide.

Design your own solutions

Every structural designer can design their own detail connections and/or reinforcements of existing containers to meet specified requirements of builders or clients. This can be a reinforcement of a column or a beam. Because of tolerances and temperature, deformation must be calculated so that the finished container building construction with units meets the local building codes and/or the specifications of the client. Bearing capacity of vertical loads and horizontal loads of each standard freight container are documented on a nameplate on the container and more detailed information can also be found in the ISO container specifications.

Adapting standard freight containers

Adaptations are mostly necessary when freight containers are used for purposes other than freight. After adaptation, the certification of a container is no longer valid with regard to its adjusted parts. The structural functionality of these altered parts must be demonstrated separately in order to ensure that the loads on the assembled building can be supported according to the local building codes. The same applies to deformation demands on the next unit, the foundation, or other structural elements outside the units.

When adapting containers—for example, removing the doors and the front side wall—a calculation of strength and deformation can be used as a way to demonstrate and guarantee that the structural unit complies with safety standards. The original parts of the container can be used as a framework with structural elements in software or hand calculations. If they do not meet the required numbers, additional structural elements as angle steel and/or box girders can be welded to reinforce the original parts in an efficient way. The requirements of the architect and the builders must still be met. The parts that are not changed keep their structural capacities or perform even better when reinforcement with additional elements is done. The changed part's stability can be demonstrated through structural calculations, and unchanged parts remain covered by ISO standards. When a sufficient number of units are changed, test loads on the changed unit according to valid ISO standards can save costs for material and calculations. The entire testing must be conducted in accordance with ISO standards or if individually organized, according to demands of the local authorities.

Conclusion

It is possible to use standard freight containers for buildings, when using the non-adapted structure capacity according to ISO specified loads as these are approved worldwide. The adaptations in the structure must be calculated for strength and deformation as standard. Testing adaptations of construction instead of calculations can often be rewarding.

This way, builders, owners and authorities can accept structural designs of buildings assembled with adapted containers as normal. Structural engineers can design and calculate this way in what is asked for the container unit and also for the building. So engineers can use in their design the advantages of these wonderful containers.

07

PROJECTS

Port-a-bach/Atelierworkshop	50
Kiowa Prototype (Habitainer)/Luis Rodríguez Alonso, Javier Presa	52
MDU/Mobile Dwelling Unit/LOT-EK	53
Bohen Foundation/LOT-EK	54
Illy Café/Adam Kalkin	56
Push Button House/Adam Kalkin	57
Containing Light Mobile Unit/EER Architectural Design	60
Dinahosting Offices/O Antidoto	62
Container Housing/Gustau Gili Galfetti	64
Uniqlo Flagship/LOT-EK	66
Uniqlo Pop-Up Stores/LOT-EK	66
Brentwood Cabana/IC Green Inc.	68
Skaeve Huse/Tempohousing/Kerssen Lijbers	69
Bornack Drop Stop Training Center/Patzner Architekten	70
Theaterhaus Stuttgart/plus+ bauplanung with engelhardt.eggler.architekten.	74
Sky is the Limit/Bureau des Mésarchitectures	76
Residential Containers/HSH Architects	78
X-Space Atelier/FPS Oficina de Arquitectura	80
Sjakket Youth Center/PLOT = JDS+BIG	81
Spaceman Spiff/Serda Architects	84
Floating House/Han Slawik	86
koelnerbox (prototype)/Jan Hohlfeld et al.	88
Fanshop of Globalization/Raumtaktik	89
Mirror Error/Yasutaka Yoshimura with Manabu Mizuno	90
Gorman Ship Shop/Nest Architects	91
HomeBox1/Han Slawik	92
Sanitary facilities for summer camp/AFF Architekten	94
Gold Pavilion/Angela Fritsch Architekten	95
c320s/HyBrid, Cargotecture	96
Holyoke Cabin/Paul Stankey, Sarah Nordby	98
Sauna Box/Castor Design	100
Future Shack/Sean Godsell	102
Infiniski Manifesto House/James & Mau Architects	104
Wijn of Water/Bijvoet Architectuur & Stadsontwerp	108
Children's Activity Centre/Phooey Architects	110
Temporary exhibition/100 years of Football Club St. Pauli/KOMMA4 Architekten, Pütz/Reetz	114

Projects

Interstitial Living / LHVH Architekten	115
PLATOON cultural development Berlin / PLATOON.Berlin & Soeren Roehrs	116
Container House (Killer House) / Ross Stevens	118
Freitag Flagship Store Zurich / Spillmann Echsle Architekten	122
Live/Work Space / Sculp(it) Architecten	124
Meetingpoint plan06 / LHVH Architekten	126
Half Mile / Leibniz University Hanover // Hilde Léon and Udo Weilacher	127
containR Cinema / Robert Duke Architect et al.	128
Eichbaumoper / raumlaborberlin	130
ORBINO / Luc Deleu	132
Construction X / Luc Deleu	134
Hoorn Bridge / Luc Deleu	134
Speybank / Luc Deleu	134
Consumer Temple – Broken Icon / Gary Deirmendjian	135
ContainerArt São Paulo / Artur Lescher, Bernardes + Jacobsen	136
GAD / MMW Architects of Norway	138
Puma City / LOT-EK	140
Suburban House Kit / Adam Kalkin	144
Old Lady House / Adam Kalkin	145
12 Container House / Adam Kalkin	148
King Kamehameha Beach Club / Angela Fritsch Architekten	152
Book Fair Pavilion / Frade Arquitectos	153
B-Camp / Helen & Hard Architekten	156
Architekturbox / She-Architekten	160
bed by night / Han Slawik	162
R4House / Luis de Garrido	164
Trinity Buoy Wharf (Container City I+II) / Nicholas Lacey and Partners	168
Keetwonen / Tempohousing / JMW Architekten	172
Raines Court / Allford Hall Monaghan Morris	174
Qubic Amsterdam / HVDN Architecten	176
Cancer Center Amsterdam / MVRDV	178
Wismar Technology and Research Center / Jean Nouvel with Ziebell & Partner	180
Campus / Han Slawik	184
Mystery Cube / Han Slawik	186
derCube / LHVH Architekten	187
seven-day architecture / LHVH Architekten	188
50,000 € Modular-House / mm+p / Meyer-Miethke & Partner	189
Temporary ACCOMMODATION for 1001 Days / Meyer en Van Schooten Architecten	190
PLATOON KUNSTHALLE Seoul / PLATOON and GRAFT Architects with U-il Architects	192
Volvo C30 Experience Pavilion / Knock	198
Nomadic Museum / Shigeru Ban Architects	200
Papertainer Museum / Shigeru Ban Architects + KACI International Inc.	204
Cruise Center / Renner Hainke Wirth Architekten	208
flyport / wolfgang latzel architekten	212
IBA_Dock / Han Slawik with IMS Ingenieurgesellschaft mbH and bof architekten	214
PIER 57 / LOT-EK	216
Billboard Building / LOT-EK	217
Sanlitun South / LOT-EK	218
Research Station / IMS Ingenieurgesellschaft mbH and bof architekten	220
Seatrain House / OMD—Office of Mobile Design	222
Zigloo Domestique / Keith Dewey	226
Parsonage Boyle Heights / DeMaria Design Associates Inc.	227
Redondo Beach House / DeMaria Design Associates Inc.	228
Chalet du Chemin Brochu / Pierre Morency Architecte	232
Hidden Valley / Marmol Radziner Prefab	234

Project
PORT-A-BACH
Architect
ATELIERWORKSHOP

Project location   Various
Estimated use   Mobile home
Container type   Freight container

01

02

03

04

Atelierworkshop fabricated its one-room Port-a-bach vacation house from a single 20-foot freight container with an additional roofed area of 14 square meters. When folded out, the inner room is supplemented with a terrace that runs along the longitudinal side of the structure and with balcony-like reclining surfaces at one of the container's open ends. Flexible furnishings can be stored in the wall. This and other efficiencies of organization mean that four people (two adults and two children) can comfortably inhabit Port-a-bach despite its minimal floor space. The house can operate in a stand-alone manner from a technical point of view, however, it can also be connected to existing public utilities (electricity, water, sewage) if necessary. The architects present Port-a-bach as a sustainable alternative to the vacation or mobile home because it can be removed without leaving any trace behind. This would suggest that entire caravan parks can be returned to nature in an easy and relatively inexpensive manner as the tourists move on to fresh destinations. Best of all, although the house is temporary and transportable on the one hand, it is also suitable for long-term and permanent use because of its robust construction based on the freight container.

01, 02
The container opens and closes easily, with beds and a deck folding or sliding outwards.
03, 04
When open, the terrace reveals the container's great space efficiency.
05, 06
The kitchen fits into a wall immediately adjacent to the fold-out bedroom.

Project
KIOWA PROTOTYPE (HABITAINER)
Architect
LUIS RODRÍGUEZ ALONSO, JAVIER PRESA

Project location : San Antonio, Canary Islands, Spain
Estimated use : Private residence
Container type : Freight container

01 – 03
The end of the container opens out as a balcony but also, when the steel walls are partly closed, serves as a doorway.
04, 05
Drawings demonstrate how containers can be fit like drawers into a larger structure and supplemented with stairs and exterior windows.

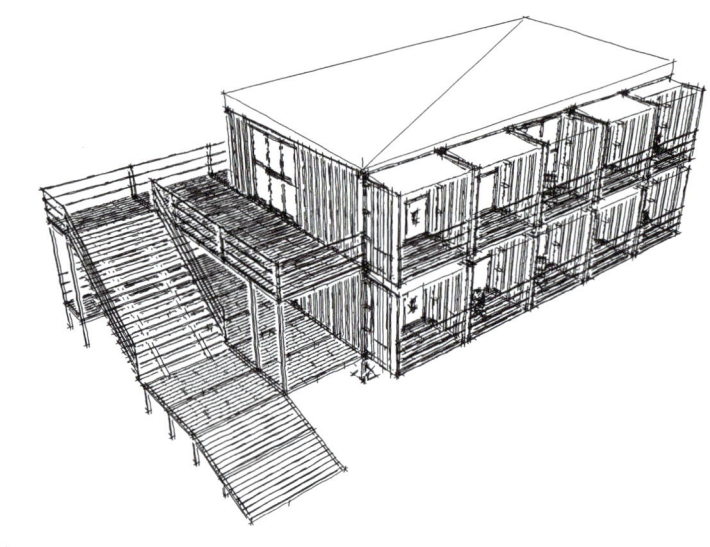

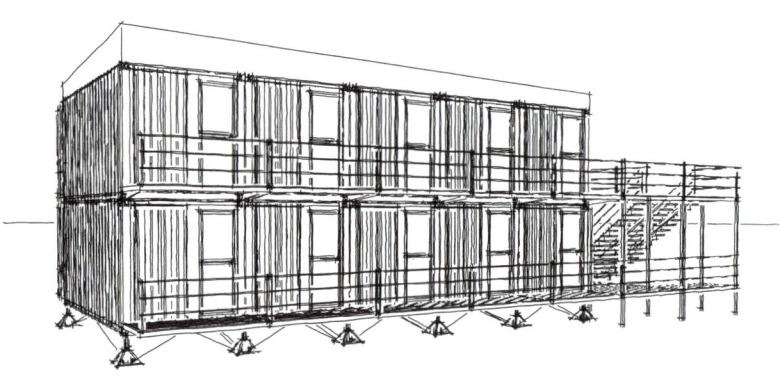

In the Canary Islands, Spanish architects Luis Rodriguez Alonso and Javier Presa created this 14-square-meter residence from a 20-foot, steel ISO freight container, which serves as the home's structural chassis. The interior panel system is a composite of polyurethane, polystyrene, wood frame and medium density recycled fiberboard. The architects painted the ceiling and walls a matte white and sheathed the floor with a seamless plastic finish, dubbing this sophisticated vessel for the containment of domestic life a Habitainer.

Project
MDU / MOBILE DWELLING UNIT
Architect
LOT-EK

Project location	Walker Art Center, Minneapolis, USA
	Museum of Art, University of Santa Barbara, California, USA
	Whitney Museum of Art, New York, USA
Estimated use	Mobile home
Container type	Freight container

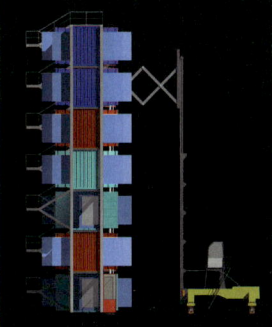

01
By cutting into the container, LOT-EK created closable drawers to expand the interior and let in light.
02
Renderings show how MDUs arriving at a port can be loaded into a track and slid into place in a community of nomadic living units.
03
An interior view.
04
Floor plans show a unit closed for shipping (top) and open for living.

This is the new Plug-In City: LOT-EK's Mobile Dwelling Unit (MDU) is created by making incisions in the metal walls of a 40-foot cargo container to define extruded subvolumes. In transit, these volumes are pushed in like drawers, allowing worldwide standardized shipping of the home. When in use, the volumes remain pushed out, leaving the interior unobstructed. Architects Ada Tolla and Giuseppe Lignano conceived of the MDU for the new nomad, enabling the residential unit to travel with its inhabitant like a vast steamer trunk, filled with the owner's belongings whilst en route. At its destination, the MDU is jacked into a multilevel steel rack by a crane that constantly loads and unloads MDUs into slots where they are anchored via steel brackets into assigned positions and plugged into local services. LOT-EK compares the units to pixels in a digital image, temporary patterns generated by the presence or absence of dwellings, reflecting the ever-changing composition of these modern-day international colonies.

Project
BOHEN FOUNDATION
Architect
LOT-EK

Project location   New York City, New York, USA
Estimated use   Office and Exhibition Space
Container type   Freight container

01 – 03
Tracks embedded in the gallery floor allow containers to be moved around the interior at will.
04
Movable walls permit views through containers and into the larger interior.
05 – 07
Graphic treatment of the corrugated steel surfaces of the containers became even more engaging when panels of each wall were folded downward to suit the artwork being displayed.

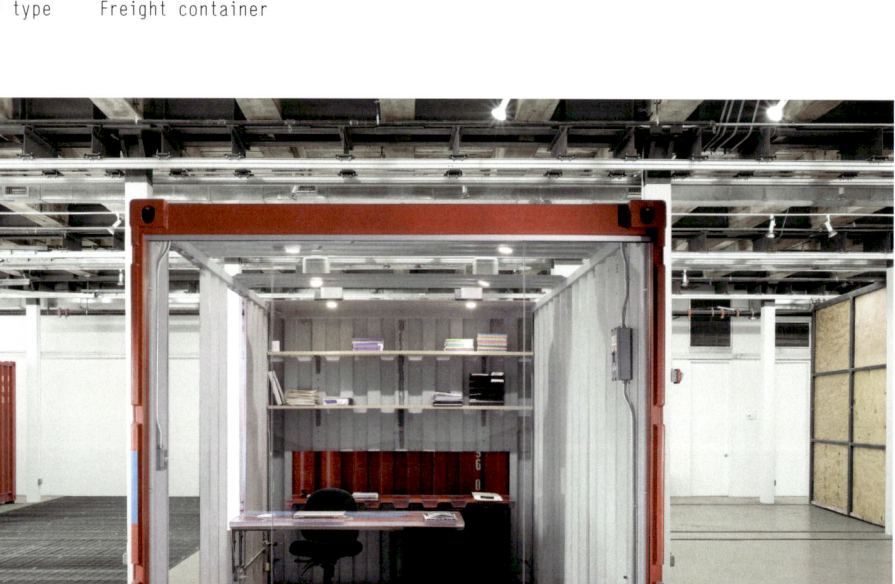

The production spaces of a former printing factory in Manhattan's Meatpacking District accommodate the gallery of the Bohen Foundation, a private foundation for contemporary art specializing in the promotion of artists whose works cannot be exhibited in "normal" gallery spaces due to their scale and complexity. This kind of multimedia exhibition concept requires maximal spatial flexibility. To meet these requirements, LOT-EK made use of eight freight containers that—tuned to the grid dimensions of the factory hall—were placed on rails embedded in the floor. The containers can now be individually moved within the venue, allowing the available interior spaces to be flexibly allocated according to the requirements of the exhibited works. The containers offer all spaces necessary to operate the gallery, including office/administration, lounges, kitchen, library etc. Since they are not needed for weather protection or to form a climatic shell, the structures of the containers were left unaltered. However, they were painted and openings of various sizes were cut into the walls and roofs. At places where an intimate atmosphere is desired, the openings were glazed; otherwise they were left open. Some of the furnishing, e.g. tables and shelves, were also cut from the walls and bent into shape.

05

06

07

Project
ILLY CAFÉ
Architect
ADAM KALKIN

Project location Venice, Italy
Estimated use   Café and Bar
Container type   Freight container

Adam Kalkin designed espresso impresario Illy's café container, which was presented for the first time at the Venice Biennale 2007. The core of the container can be converted into a café, where guests may sit at a long table with a small library of books under a chandelier. The bar, with no less than three espresso machines, anchors one end of the container, and also conceals all the electrical wiring. To the left, a small workplace includes a desk and computer, and is adjoined by a washroom. To the right, the lounge consists of a sofa and side tables. The bright white interior was accessorized with a bright white awning.

01
Illy's coffee bar in the process of folding open via hydraulic mechanisms.
02
The bar closed after hours.
03
The bar fully open in the literal and architectural senses.

Project
PUSH BUTTON HOUSE
Architect
ADAM KALKIN

Project location   Various
Estimated use   Mobile home
Container type   Freight container

01

04

05

02

03

01 – 03
A view from the long side as Push Button house unfolds automatically—and in under one minute.
04, 05
View from the end of the container into the house as its walls fold open.
06 [pages 58–59]
When open, the dwelling features five rooms: kitchen, dining and living room, bedroom and library.

Just like the Illy Café, Kalkin's Push Button House unfolds like a flower at the push of a button. Here, the closed ISO freight container transforms into an single-unit home including kitchen, dining room, bedroom, living room, and library, by unfolding hydraulically. Beyond it's exemplary nature as a mobile home, the Push Button House became an artwork itself due to it's high level of transformation. In December 2007, Push Button House was exhibited in New York City where it welcomed holiday shoppers at the Time Warner Center. As part of the New York Wine & Food Festival, Push Button House again took center stage in the Meatpacking District in October 2008.

Project
CONTAINING LIGHT MOBILE UNIT
Architect
EER ARCHITECTURAL DESIGN
GEERT BUELENS AND VEERLE VANDERLINDEN

Project location	Various
Estimated use	Showroom
Container type	Freight container

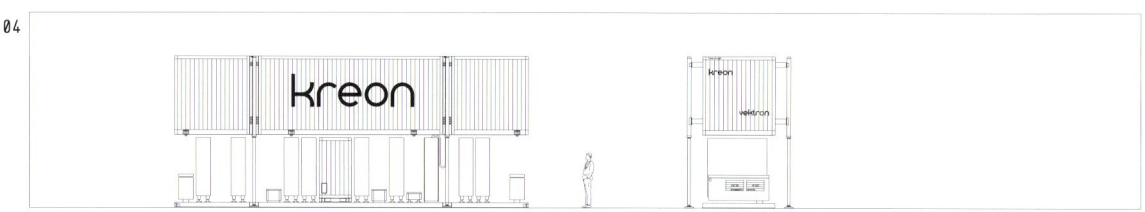

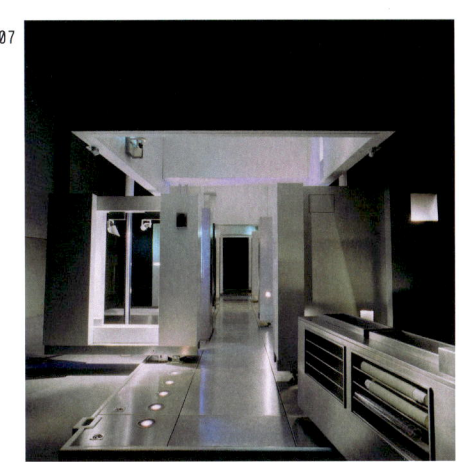

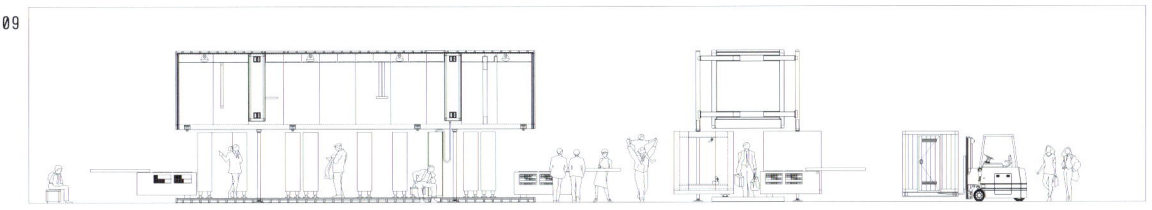
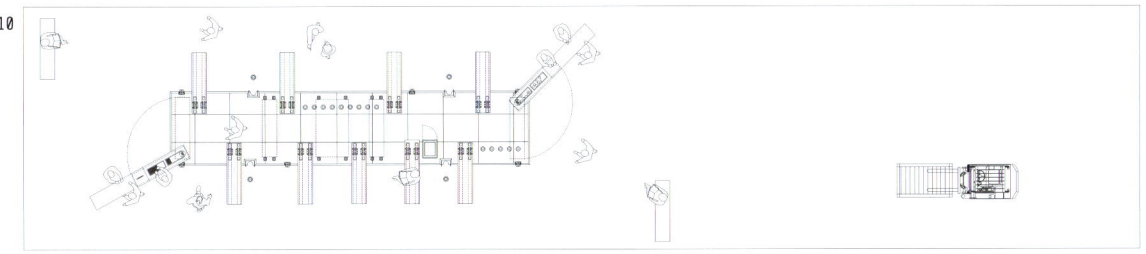

Kreon, the architectural lighting company that introduced the first semi-prefabricated exhibition stand, commissioned Eer to make a trade fair booth for Euroluce promoting the brand and to improve on its first efforts. Their mobile exhibition container, in which everything is pre-wired and integrated into the architecture, which is no longer static, but dynamic and "containing light". A truck transported the 40-foot container to the exhibition ground where it was jacked up by four five-meter high computer-controlled hydraulic cylinders, which were already integrated into the container. Having been packed with its own exhibition cabinets, when these were removed and put into place for display, the volume of the original container was doubled and a rather small initial volume was converted into a spectacular architecture. The result was also a fine demonstration of how architecture can be defined and shaped by light. While Eer kept the external shell of the container largely as-is, they refined the interior by using highly polished stainless steel and white leather to create a less industrial atmosphere.

01
Kreon's mobile showroom closed.
02
The showroom ready for transport by truck.
03
The container contained four hydraulic cylinders on which it could be jacked up on arriving at the fair grounds.
04, 05
Views in cross section (top) and floor plan (bottom).
06, 07
Interior seen from side and end.
08
Open for business, a desk swings outward.
09, 10
The booth open, views in cross section and floor plan.

Project
DINAHOSTING OFFICES
Architect
O ANTIDOTO
ANTONIA MAIO AND JAVIER QUINTEIRO

Project location   Santiago de Compostela, Spain
Estimated use   Office space
Container type   Freight container

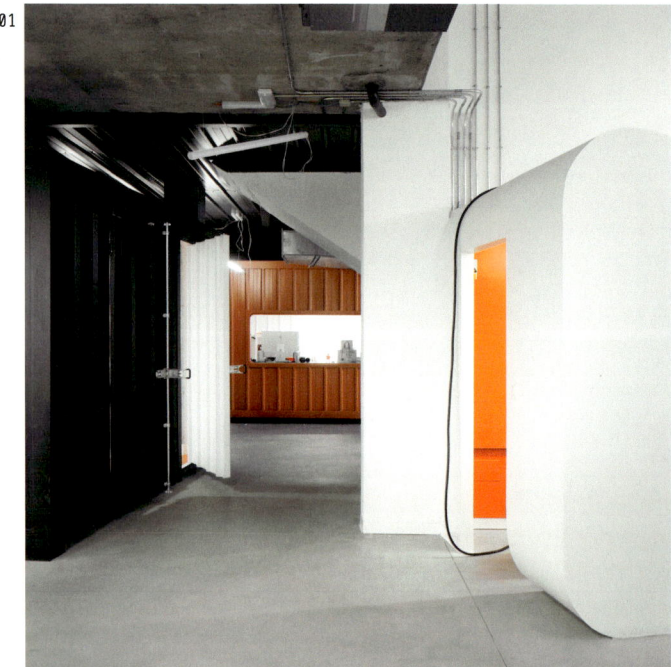

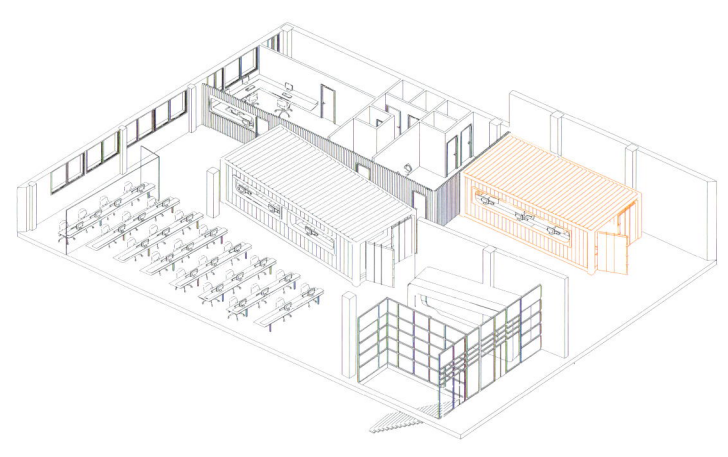

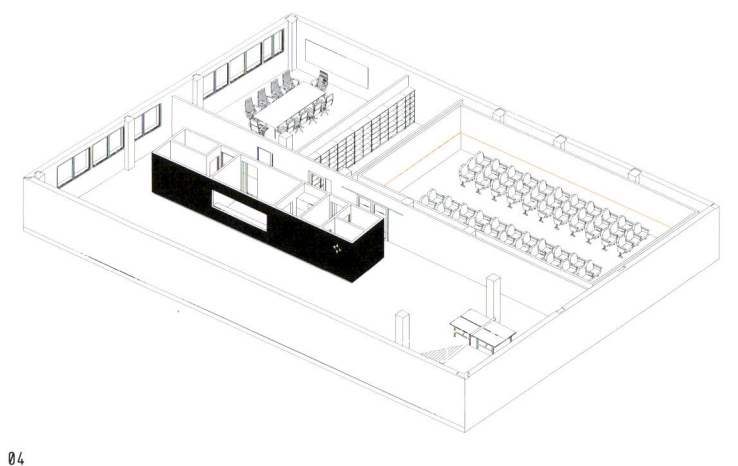

01, 02
The interiors of this new office were supplemented with containers.
03
Ground floor.
04
Basement.
05 [page 63]
Workspace seen from the container exterior.

A former two-story gymnasium in Santiago de Compostela was chosen for creative agency Dinahosting's new 885-square-meter headquarters and offices. O Antidoto designers Antonia Maio and Javier Quinteiro placed offices and working space on the ground floor and common areas, auditorium, boardroom, and exhibition room/cafeteria in the basement. By using industrial containers, the designers echoed the innovation, international youthfulness, and aesthetic of the agency while providing a durable material capable of sustaining heavy use, as well as giving the project a distinctive personality. Given the irregular distribution of the pillars and columns in the existing space, along with the need to open the facade to bring the recycled shipping containers inside, the team decided instead to assemble them on the site. This approach suggested the use of corrugated steel as a finish in several areas. The resulting workspace resembles a tidy warehouse distinguished by an area filled with piled up containers that hold private offices and toilets. These boxes provide counterpoint to an open space devoted to phone support, reception and circulation. The reception area was conceived as an independent section, the softer lines of which contrast nicely with the severity that characterizes the bulk of the interior: it consists of a graphical contour of Pladur (plasterboard) lacquered in white and corporate orange.

Project
CONTAINER HOUSING
Architect
GUSTAU GILI GALFETTI

Project location   Barcelona, Spain
Estimated use   Housing system (prototype)
Container type   Freight container

Built by Catalan architect Gustau Gili Galfetti, this temporary exhibition prototype for a generic housing system relies on the marriage of two clearly distinct, but joint, domestic spaces. The first space feels unfinished, open, neutral, and indeterminate—easily "susceptible" to the influence of its inhabitants. The second environment is compact, closed, technological, and prefabricated. In order to generate this drastic dichotomy in a home, Galfetti used a modular structure of reinforced concrete pillars, complimented by an assemblage of recycled ISO 20-foot maritime containers. Galfetti did not relegate domestic space to the interior of the containers only; instead he created home inside, alongside, above, and around them, amplifying his material's capacity to contain, as well as to define space outside, as well.

01, 02
Renderings of the container towers in the Barcelona city fabric and standing alone.
03
The container is also visible from the building corridor.
04
Interior view.
05, 06
Galfetti's exhibition prototype.

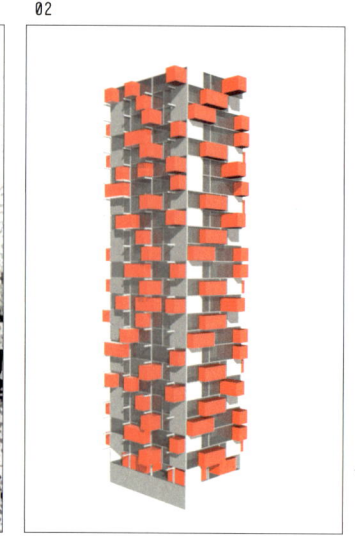

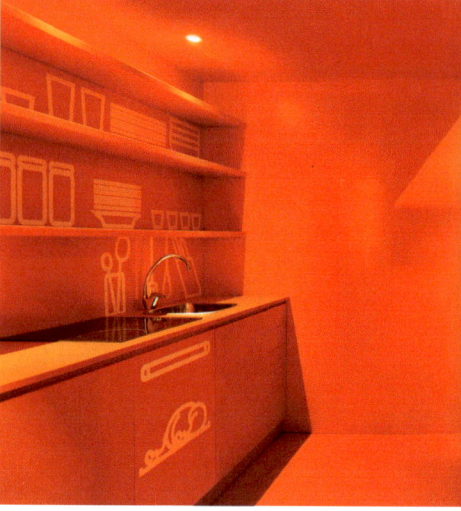

Project
UNIQLO FLAGSHIP, UNIQLO POP-UP STORES
Architect
LOT-EK

Project location	Osaka, Japan
	New York City, New York, USA
Estimated use	Retail space
Container type	Freight container

01
Exterior of the Osaka flagship.
02
A shop transported by truck.
03
Interior of a New York City shop.
04
The exterior of a retail container in the Soho neighborhood.

01

02

03

For over a decade, the work of New York firm LOT-EK has exemplified the creative reuse of shipping containers, truck parts and tanks, along with other icons of the industrial world. For Japanese clothing retailer Uniqlo, the studio designed a 1,022-square-meter flagship in Osaka, Japan (that incorporated a single freight container into the building), along with pop-up shops in New York City. The pop-ups comprised two 20-foot containers that toured Manhattan to introduce the brand to the United States. Traveling on a knuckle-boom truck, these containers functioned as retail space, fully equipped with shelving, cash-wrap, and fitting room. Vertical windows aligned with shelving displays folded and anchored merchandise to the exterior. The fitting room was a custom-designed duct that gathered like an accordion towards the ceiling via a manually cranked pulley system. Linear bands of light illuminated the interior from both floor and ceiling while dressing mirrors lined both ends of the containers in order to visually elongate the limited space.

Project
BRENTWOOD CABANA
Architect
IC GREEN INC.

Project location   Los Angeles, California, USA
Estimated use   Pool house
Container type   Freight container

This 30-square-meter cabana was installed in a tony Brentwood backyard by IC Green, a design-build company that delivers fully engineered, affordable structures made from reclaimed shipping containers. The structural rigidity of the metal frame and its corrugation, rival standard timber and steel construction (in strength and build time), as well as being easy to stack and cluster in various configurations. It also allows for more square footage within the same footprint as a standard timber construction. The cabana will also consume less energy through the integration of passive and active systems, water conservation measures and eco-friendly materials.

01, 02
Views into the cabana.
03, 04
Interior details, including bench and shower.

Project
SKAEVE HUSE
Architect
TEMPOHOUSING / KERSSEN LIJBERS

Project location   Amsterdam, the Netherlands
Estimated use   Social housing
Container type   Freight container

Tempohousing's 180-square-meter social housing consists of six salvaged shipping containers with specially adapted front and rear frames which required no super-structure. On the ground floor, the containers are mounted on Stelcon plates and faced with stone and concrete. The temperature in each container may be adjusted individually via mechanical ventilation with automatic variable speeds, while a four-group fuse box provides 50 litres of hot water per container. Each interior is equipped with an electrical stove, a bathroom with shower, toilet and sink, and a full kitchen.

01 – 03
Exterior views of six units of social housing which were staggered, clad with stone and concrete and set on Stelcon plates.

Project
BORNACK DROP STOP TRAINING CENTER
Architect
PATZNER ARCHITEKTEN

Project location	Marbach am Neckar, Germany
Estimated use	Training center and office space
Container type	Building container

01

02

For anyone—from elevator and bridge repairers and outdoor adventurers, to rescue squads and construction workers—Bornack offers bespoke security systems, coaching and consultation, as well as "drop-stop" safety equipment. Patzner Architects transformed a monumental, landmarked former steam power plant in Marbach am Neckar (built in 1939 and closed in1951) into its fourth training and product development center. To do so, the designers tore out all the old coal boilers except one inside the boiler house, adding 500 square meters of space over several floors to this industrial cathedral of 3,000 square meters. Patzner converted the factory into offices, training ranges, camps, and meeting and break rooms while working hard to reconcile the magnificent original structure with these new functions.

01
Containers added more intimate space, for safety training, to this former steam power plant.
02
The new, but still industrial, stairwell.
03
The end of each container was fully glazed to give monumental views of the building's main volume.

By raising shipping containers 6.5 meters off the floor, the designers underscored the church-like volume and epic 35-meter ceiling height of the existing building (this is also emphasized by the new stairs which hang from two powerful old steel girders). By completely glazing the ends of the boxes, the architects punctuated the space with new layers of texture and transparency. Inside the containers, visitors seem to float in midair while enjoying a magnificent view of "the mountain climbers", clients who train in drop-stop safety directly opposite these windows. Patzner was wise to preserve the existing building strategically and generously, including rehabilitating concrete and steel surfaces, in order to communicate the original character and surreality of the plant environment.

04  05

04, 05
The windowed containers hem one flank of the factory, accessed via stairs, and give views onto training sessions taking place in the larger volume of the space.

Project
THEATERHAUS STUTTGART
Architect
PLUS+ BAUPLANUNG
WITH ENGELHARDT.EGGLER.ARCHITEKTEN

Project location   Stuttgart, Germany
Estimated use   Event space
Container type   Freight container

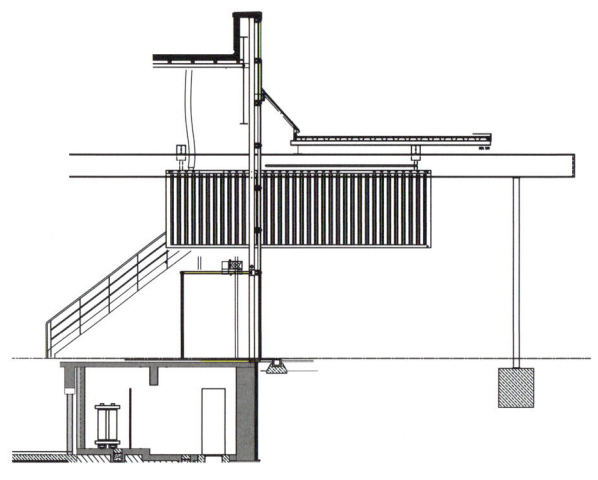

This architectural reinvention puts a modern face on an historic landmark factory building, connecting two periods in time without disjunction. The architects suspended a freight container just above the main entrance that looks out onto the adjoining plaza, serving as a fine advert for the theater and creating a single windowed room. The energy-saving extension embraces the old structure (literally) while giving it both fresh purpose and authenticity.

01
Visible at the entrance: a container used in the renovation and the modern cladding placed over the old brick facade.
02
The entrance seen from inside.
03
The facade seen in cross section.
04
The facade seen from the plaza.

Project
SKY IS THE LIMIT
Architect
BUREAU DES MÉSARCHITECTURES
DIDIER FIUZA FAUSTINO

Project location   Yangyang, South Korea
Estimated use   Tea house
Container type   Container frames

01, 02
The tea pavilion above Yangyang.
03
Inside one of the pavilion's twin containers, looking toward the Sea of Japan.
04
The stairwell seen from one of the container-rooms.
05 [page 77]
The containers hang 20 meters above ground.

Along the shoreline in Yangyang, South Korea, Sky Is The Limit is a 100-square-meter tea pavilion, suspended 20 meters above the ground overlooking the Sea of Japan. Didier Fiuza Faustino's Paris-based Bureau des Mésarchitecture projected this steel-framed tearoom above the horizon and lined it with varnished plywood, glass and steel duckboard. The building's body resembles a curious robotic skeleton, a Japanese Wall-E, its thin tension-rod laced structure accessed by a series of staircases that lead visitors up to the creature's "eyes," a pair of generously open-ended viewing platforms made from container frames. The two voids, of identical dimensions, provide diametrically opposing views of the landscape below.

Project
RESIDENTIAL CONTAINERS
Architect
HSH ARCHITECTS

Project location	Prague, Czech Republic
Estimated use	Private residence
Container type	Container frames

01

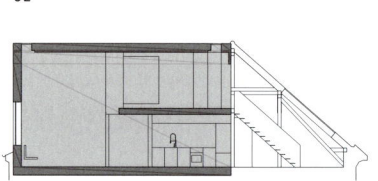

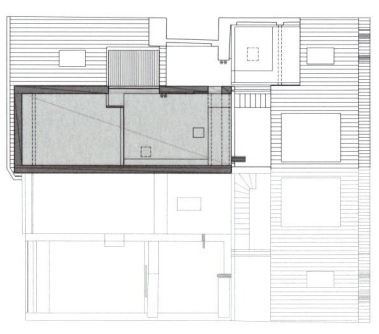

01
Structural boxes crown this otherwise bland building.
02
Section drawing and floor plans.
03 – 06
Details of the interior showing the colour scheme.
07
Axonometric drawings showing the insertion of structural boxes and how their walls can be folded to create more efficient volumes.

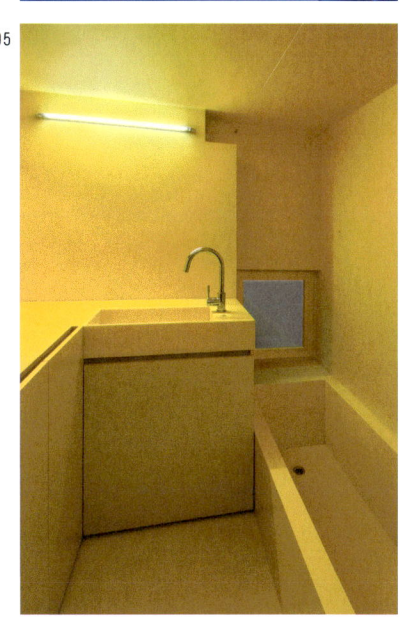

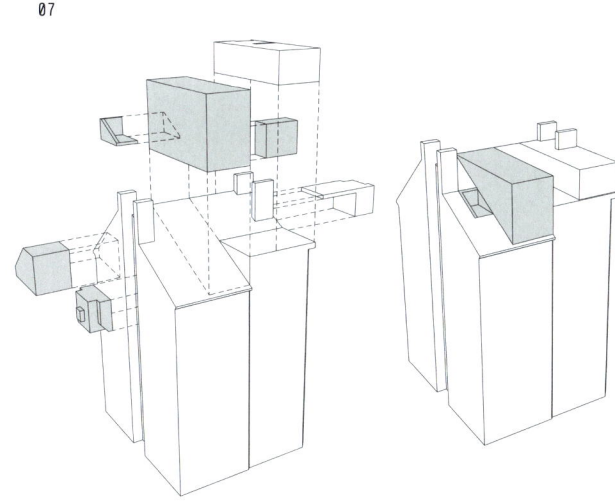

This Prague-based project consists of structural boxes inserted into an existing roof in a way that does not affect the readability of the original building. The result? Residential spaces that, at 152-square-meters, contain living rooms, restrooms and technical facilities with bedrooms and cloakrooms tucked into the original attic. It is possible to close the inserted boxes by folding the interior walls to create more compact volumes. Large openings, which the HSH Architects call "the eyes", have been cut into the walls to give a view out, but can just as well remain shut. All structures were constructed using common building technologies and materials (timber, glass, metal, plasterboard). To celebrate the difference between the old and new structures, the roughness of the original roof was preserved and no surface treatment applied. By contrast, the newly inserted components were clad in smooth sheets of titan zinc or painted in monolithic and vivid planes of color.

Project
X-SPACE ATELIER
Architect
FPS OFICINA DE ARQUITECTURA
FRANCISCO FENILI, JORGE PÉREZ GONZÁLEZ AND JULIO SEPIURKA

Project location Buenos Aires, Argentina
Estimated use   Atelier
Container type   Container look

01
Access to the artist's studio is via exterior stairs.
02
One end of the container features floor-to-ceiling glazing and great views of the city.
03
The aritst at the door of her workspace.
04
Florr plan of the container-based extension.

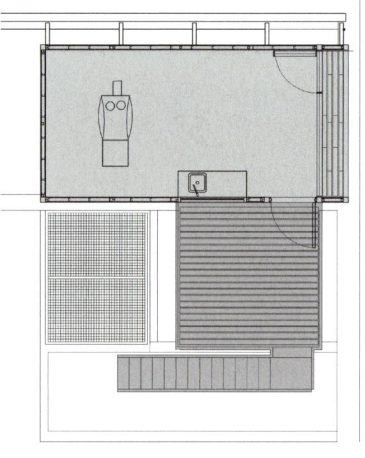

In an old section of downtown Buenos Aires, a textile designer lives in a building that integrates her life and work, home and studio. The challenge for Argentine architects FPS in this extension project was to create a new space housing the atelier that would be inexpensive and would not spill into the rest of the house. The designers decided to add a rooftop extension in the form of a steel container, a choice that resulted in a striking and rich contrast between the old masonry dwelling and the fresh metal workspace. Inside, the artist now enjoys views over the city through large floor-to-ceiling windows. This gives her the impression of being safely tucked away in an urban refuge while still being able to explore the inspiring living creature that is San Telmo.

Project
SJAKKET YOUTH CENTRE
Architect
PLOT = JDS + BIG

Project location   Copenhagen, Denmark
Estimated use   Community building
Container type   Container look

JDS morphed a disused and mold-infested factory into a 2,000-square-meter cultural centre for Copenhagen's immigrant youths. Local regulations dictated that the factory's twin-peak silhouette and gables be preserved, so the architects converted the existing structure. They gutted one of the vaulted buildings to create a sports hall and then tucked the smaller functions of the facility into the second half. A large garage door installation allows its south flank to open into a courtyard. The building was designed to "speak the language of the streets" and the language of its users, so the architects decided not to remove graffiti on the exterior, but to instead use the work as inspiration for the building's color scheme. The existing raw industrial architecture was also preserved and emulated in the only architectural addition to the building: a long narrow block, clad in corrugated metal and painted red, which sits on the roof of the existing factory building. Positioned at an angle, it echoes the containers in the nearby port, houses the studio of Ghetto Noize Records and has become a strong visual symbol of Sjakket's presence on the industrial skyline of northwest Copenhagen.

01, 02
A structural box crowns the new youth center and houses a music label.
03
The skyline seen from the music offices.

- 81 -

04
A path on the roof; the corrugated steel box recalls containers in the nearby port.
05, 10
Existing graffiti determined the colour scheme.

06, 07
Skylights and large glass wall panels bring ample light inside.
08
Roof with green spaces and box.
09
Cross section of the building.

Project
SPACEMAN SPIFF
Architect
SERDA ARCHITECTS

Project location London, Great Britain
Estimated use   Residential
Container type   Building container / freight container

01
Serda's renovated barge.
02
The container used to add domestic space on the boat.
03
Interior detail.
04
Views through the newly glazed container walls.

In the London district of Chelsea, directly under Albert Bridge, there's an inconspicuous old barge, the current usage of which can only be discerned when taking a look at its neighborhood that boasts white, elegant houseboats and yachts—for its outer appearance hardly changed after it was converted from a freight ship into a houseboat. Austrian architect Eduard Serda bought the inland freight ship, constructed in the 1960s, and converted it into a spacious, 250-square-meter residential ship in just five months. The core of the ship is the former freight compartment, which he transformed into a spacious living room. Its ample space is enhanced by a roof that can be fully opened—a result of its original use as a cargo compartment. One freight container and two smaller prefab containers were installed in the cargo compartment to subdivide the space in a functional way. All three containers can be hydraulically lifted to the deck, where they are then living spaces that can be opened to the outside, or allow the entire living area to be used for other purposes.

05
Serda used building containers to create his floating bachelor pad.
06, 09
The space into which bedroom containers slide below decks for storage.
07
A room lined with Japanese screens.
08
Sisal flooring mixes with outrageously floral upholstery.

Project
FLOATING HOUSE
Architect
HAN SLAWIK

Project location   Lausitzer Seen, Germany
Estimated use   Private residence
Container type   Container frames

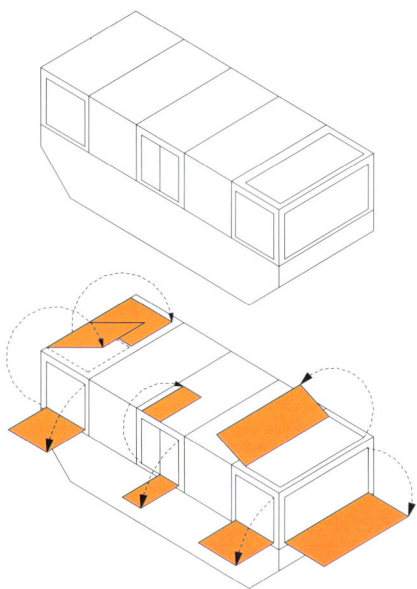

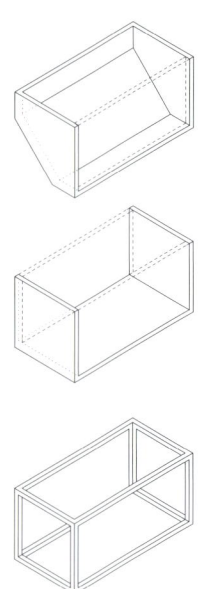

01

02

01
Passive state: structure closed, active state: wall and ceiling surfaces open.
02
Modular concrete floats and light container frame.
03
Open interior space.
04
Boxes as partitions.
05
Closed in passive state and open when in use.
06
Cross section.
07
Longitudinal section.

This prototype of a floating house proposes a container-frame building method based on a modular principle, allowing the construction of different sized houses. The modular building components are especially suited for prefabrication. A floating body made of reinforced concrete guarantees an optimal center-of-gravity position, while the superstructure is made of wood or steel in lightweight construction. The floating body consists of transportable concrete modules that are connected to form a rigid floating foundation. The advantages are high stability and maintenance freedom. The superstructure made of supporting modules, on the other hand, are lightweight wooden or steel constructions. To achieve more freedom in regard to production, usage and future alterations, the supporting and filling building components are separated from each other. Due to the modular approach, a variety of building structures can be configured. The floor plans can be adapted to the lifestyles of its users: Numerous variants are possible, from the holiday house "Mini" to the child-friendly residence "Maxi". The non-supporting, dividing building components can be prefabricated according to a uniform modular system and assembled, disassembled, and reassembled with the help of a simple click system.

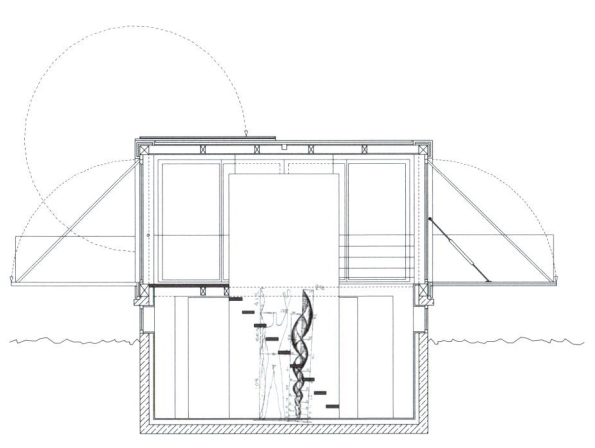
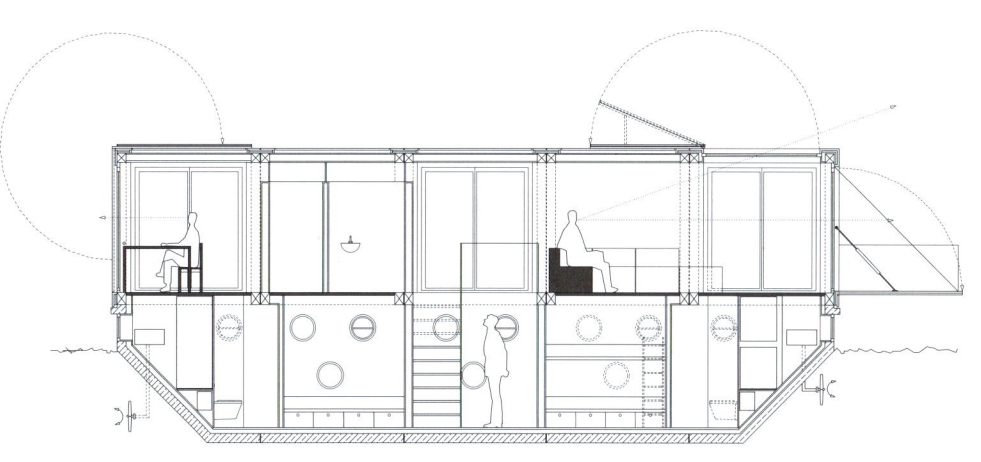

Project
KOELNERBOX (PROTOTYPE)
Architect
JAN HOHLFELD, CLAUS HESEMANN, MARKO HEINSDORFF
WITH HERIBERT WEEGEN, UWE HARZER AND ANNE MEYER

Project location   Cologne, Germany
Estimated use   Exhibition space
Container type   Freight container

This sculpture is also a prototype presentation and communications room inside a 12-meter-long High Cube (HC) shipping container. The converted interior of the container consists of wall panels and cubes arranged in a modular, meandering way and fitted with lighting arrays. These panels and boxes appear to grow out of the floor, walls and ceiling and can be rearranged and replaced by other elements if so wished. Boxes with doors attached to them provide storage space, while other boxes clad with glossed plastic and glass serve as lamps. The outer shell and the structural elements of the shipping container were essentially retained, but the four narrow windows on the side walls, and a door at one end, open the container to the exterior. At night, the koelnerbox becomes a walk-though artwork, an illuminated sculpture.

01
A view into the prototype artwork in an HC freight container.
02
The exterior was preserved except for the addition of four vertical windows.
03
Inside, modular panels, boxes and lighting turned the container into sculpture.

Project
FANSHOP OF GLOBALIZATION
Architect
RAUMTAKTIK
FRIEDRICH VON BORRIES AND MATTHIAS BÖTTGER

Project location   World Cup host cities 2006, Germany/Austria
Estimated use   Exhibition space
Container type   Freight container

Raumtaktik's traveling exhibition was designed by architects Matthias Böttger and Friedrich von Borries to mark the FIFA World Cup. The German Federal Center for Political Education was trying to use soccer as a starting point to draw attention to the economic, political and cultural developments associated with globalization: the five areas covered by the exhibition were interconnection, value creation, migration, cultural identity, and social segregation. During the event itself, the exhibition was transported in a converted shipping container through the host cities in Germany, and could also be seen in the Museum for Working Worlds in Austria.

01
A traveling exhibition in a converted container for the FIFA World Cup.
02, 03
Interior views.

Project
MIRROR ERROR
Architect
YASUTAKA YOSHIMURA
WITH MANABU MIZUNO

Project location	Tokyo, Japan
Estimated use	Showroom
Container type	Freight container

01 Exterior view of the Mirror Error installation.
02 Interior view.

Yoshimura's temporary Mirror Error installation in Shinjuku for Tokyo's Designers Week, was a 30-square-meter booth made from a single 40-foot ISO shipping container for a mail-order furniture catalog. Containers that are just about to be disposed of cost less than those that can be repurposed, so the artist worked with an old container, making countless holes in it. "Our goal was to achieve a simple subtraction design" he explains. As visitors looked through the holes, furniture showcased in the catalog could be seen ranged about the interior. Each piece was fabricated slightly differently from the others, and guests were invited to try to identify the nature of the design variations and then browse the catalog to find out if they were correct. The trick worked: visitors to the installation ended up thumbing through the catalog extensively.

Project
GORMAN SHIP SHOP

Architect
NEST ARCHITECTS

Project location Australia
Estimated use   Retail space
Container type   Freight container

01
The exterior of the shop.
02
Nest alluded to the container's original function by retaining the raw look of its surfaces.
03
Tiny windows cut into the walls could be closed during closing hours.

Nest created a portable and sustainable 14.5-square-meter retail outlet to promote the Gorman Organic brand while touring Australia. Instead of a cash register, fancy carry bags and discount bins, this little red shipping and shopping container offered a laptop terminal for online purchasing. The rawness of the interior and the designers' deliberately minimal intervention evoked the container's previous purpose. In addition, all the fit-out materials were either recycled or collected from sustainable sources and designed to sheet sizes to minimize waste.

Project
HOMEBOX1
Architect
HAN SLAWIK

Project location   Various
Estimated use   Private residence
Container type   Building container

01
The movable home is fitted with steel corners for easy shipping.
02
Looking down the stairs.
03
In HomeBox1, floors are of equal heights.
04
Asymmetrical stairs make the most of very little space.
05
The kitchen.
06
View from the rear.
07
Slawik's highly efficient floor plan.

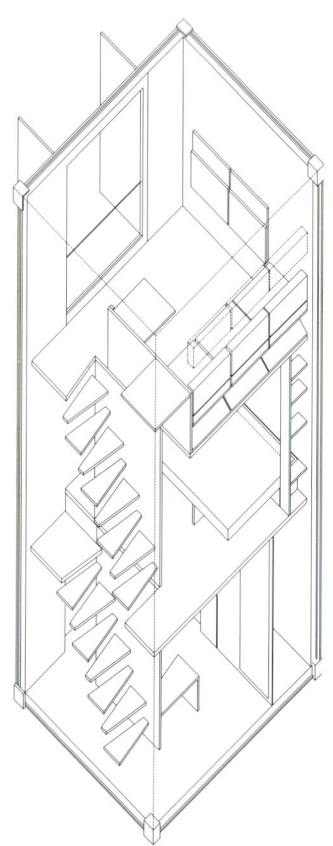

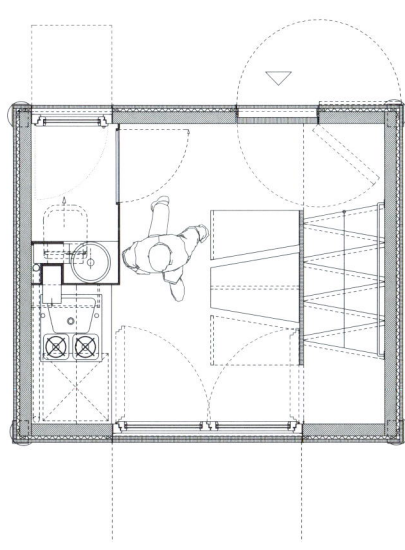

Professor Han Slawik's HomeBox shelters a miniature mobile house in a wooden container instead of the typical steel boxes in wide use around the world. Depreciation, repair and maintenance are all more expensive for steel (in terms of purchasing costs, welding, etc.) than for wood. A vertically positioned HomeBox requires little surface area and stands like a "nomad" on urban infill or rural sites and while clustered, through its repetition can create a new species of city fabric (a lane, path, street, or square). Slawik suggests that the dwelling could be used as emergency shelter or even, during major temporary events such as the soccer World Cup, an Expo or Olympic Games, erected as temporary container cities or villages to provide hotel facilities and lodgings. HomeBox1 has three levels, all of equal height, whereas HomeBox2 has larger upper and lower levels; the mezzanine has a lower height and is intended as a sleeping level. The containers are designed to be independent of location, and feature a closed rear with a generously opened face. The supporting walls and ceilings consist of multilayered wood panels equipped with steel ISO corner fittings so that they can be transported by truck, train, and ship and set up using standard lifting devices such as cranes. The corners are screwed to the wooden walls with welded, perforated panels. Slawik's design ensures that not only the corners but even a single panel can support the entire box.

Project
SANITARY FACILITIES FOR SUMMER CAMP
Architect
AFF ARCHITEKTEN

Project location   Magdeburg, Germany
Estimated use   Sanitary facilities
Container type   Building container

01, 03
The entrances to the facilities.
02
View from the rear.
04
Simple sterile surfaces were complimented on the interior with textured finishes.
05
The units seen in floor plan.

The youth camp, founded in the 1930's, runs mainly during the summer months, lodging guests in bungalows or tents. It needed temporary sanitation facilities with toilet and shower areas. To accommodate a contracted timeline for design and construction, AFF architects decided to use prefabricated cells so that the addition would be characterized by an industrial image, while extending the building containers with inclined roof dormers and vestibules that featured a simultaneously repetitive and varied pattern. Inside, the facilities were dressed with a textured surface that breaks up the immaculate whiteness of the sterile hygienic surfaces.

Along with the use of the modules, the choice of color scheme helped to create the impression of an unconventional building that was built with the use of minimal tools and which allows the original building components to remain plainly visible.

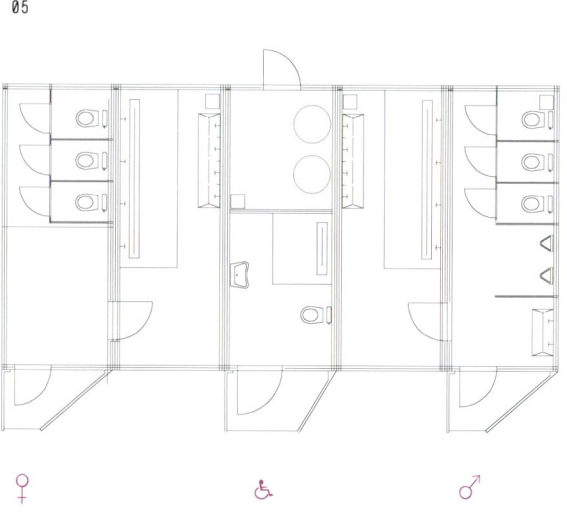

Project
GOLD PAVILION
Architect
ANGELA FRITSCH ARCHITEKTEN

Project location Darmstadt, Germany
Estimated use   Health center
Container type   Container frames

01

Commissioned to create a Holistic Health Center—a sanctuary for alternative therapies including acupuncture and acupressure—for the Alice Hospital in Darmstadt, local architect Angela Fritsch designed a 48-square-meter space that incorporated one container frame. The golden box was situated in a park adjacent to the hospital and filigreed with leaf-like cut-outs that provided full window openings as well as smaller, uneven apertures that brought light into the space as if it were pouring through a canopy of trees.

01
The finely graphical cladding lets light penetrate in delicate patterns.
02
The entrance to the pavilion.
03
A floor plan of the health sanctuary.

02

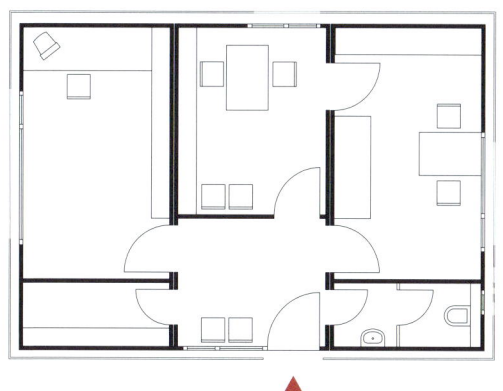

03

Project
C320S
Architect
HYBRID, CARGOTÉCTURE

Project location — Enumclaw, Washington, USA
Estimated use — Private residence
Container type — Freight container

A fine example of cargotecture, Studio 320 is an off-the-grid, low-impact living space that can be used in either an urban or rural setting. The c320s owner may relocate the entire 30-square-meter home in a single day, if desired. The residence consists of a great room that takes up three-quarters of the interior, plus a separate bathroom and sleeping area. Since 2003, HyBrid has focused on prefabricated architecture, pulling together teams of engineers, suppliers, fabricators and systems partners to design and build the most cost-efficient and sustainable planned prefab projects. The c320s features two 20-foot containers that are staggered past one another by 6 feet. In an urban environment, the units can be stacked three high, with an adjacent exterior stair, a solution intended to address the densification of neighbourhoods that were previously zoned for single-family residences. Its foundation is constructed from a few pre-cast concrete footings, and the roof is seeded to grow into an intensive green fern-based roof. HyBrid selected propane to supplement ecological energy sources in order to sustain itself off-the-grid. The bathroom is tied to an above-ground "green machine" septic system, but could also feed into a conventional system. The structure is fully insulated to international code and its interiors are designed to last without off-gassing; walls, floor and ceiling are built of durable finish plywood, fixtures are stainless steel, and counters are waterproof paperstone. HyBrid treated the project as a living laboratory for adding self-sustainable features resulting in a home that features a water-collecting green roof, extensive recycled materials and the potential for habitat restoration.

03

04

05

06

07

08

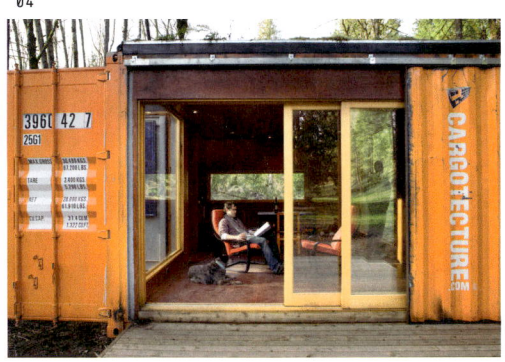

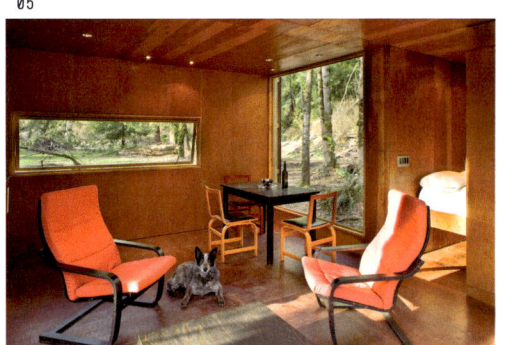

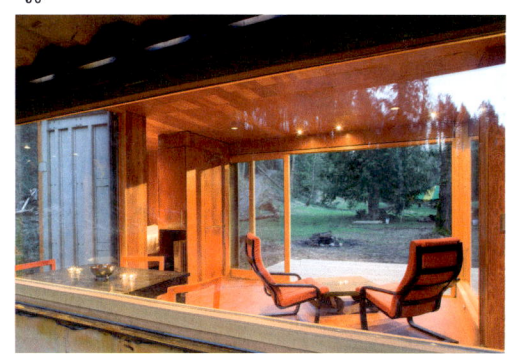

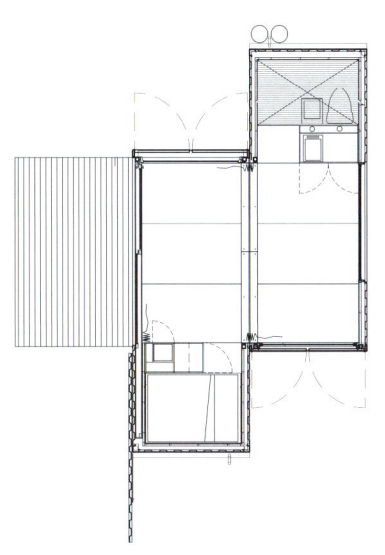

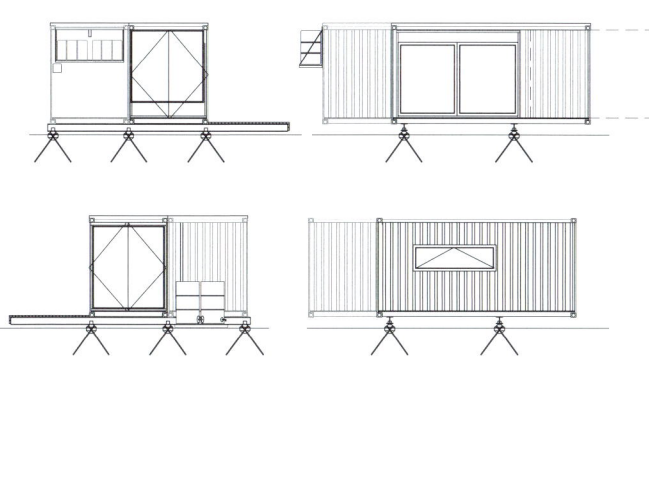

01, 02
The two-container dwelling open and shut.
03
Solar panels point to the home's many ecofriendly characteristics.
04
Opening the home reveals generous glazing all around.
05, 06
Interior details.
07, 08
The living-sleeping container is offset slightly from the bathroom container. The boxes rest on pre-cast concrete footings.

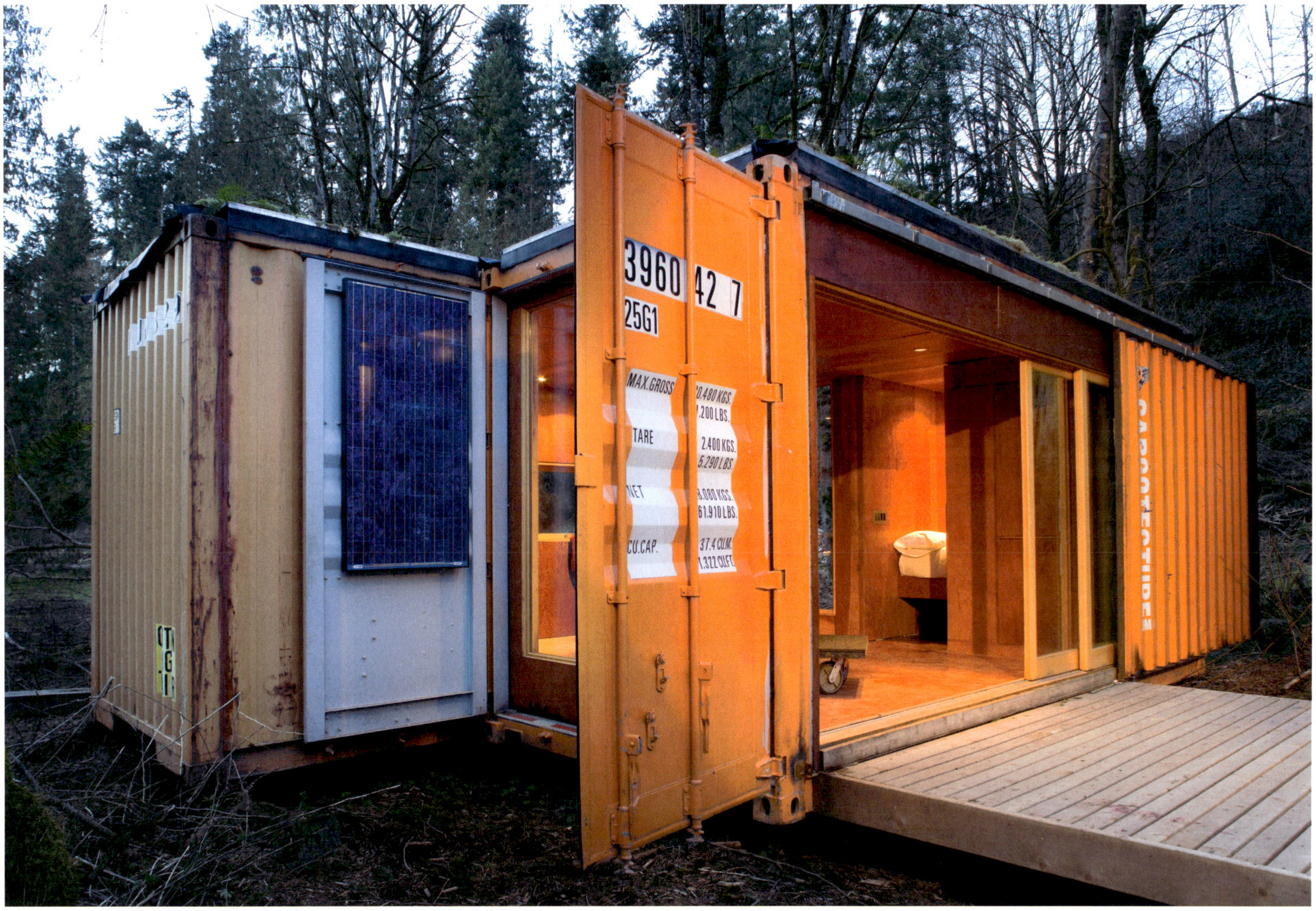

Project
HOLYOKE CABIN
Architect
PAUL STANKEY, SARAH NORDBY

Project location Holyoke, Minnesota, USA
Estimated use   Private residence
Container type   Freight container

01
The building and its components.
02, 03
A pair of containers flank a central roofed timber construction.
04
A window in one of the containers relates that volume to the heavily-windowed core structure.
05, 06
Interior details.
07
Each container has its own foundation.

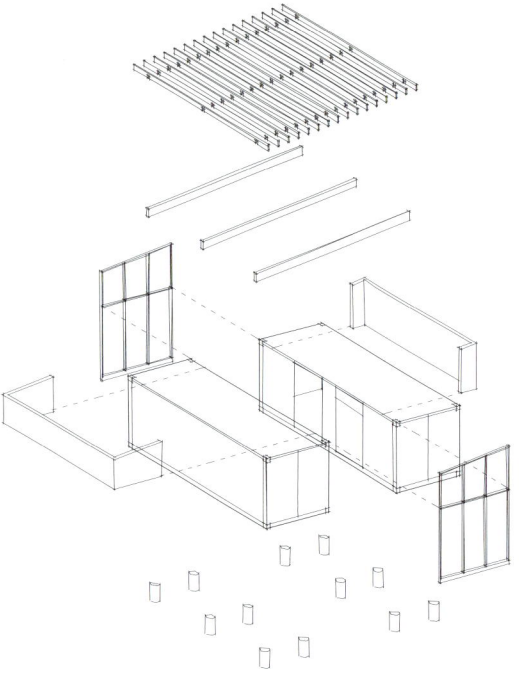

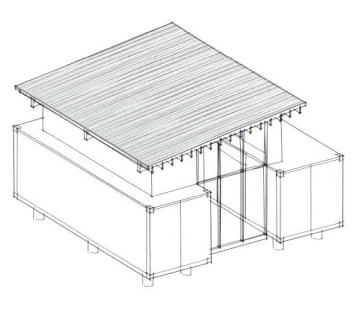

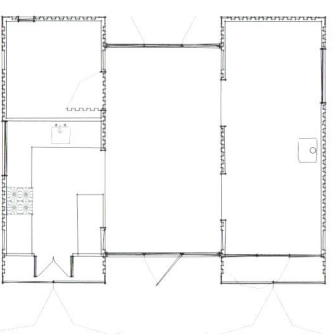

This weekend home in the small town of Holyoke, Minnesota, was built by Paul Stankey and Sarah Nordby, the couple wanted to replace a trailer home that had fallen into disrepair. The architects' goal was to come up with as permanent and robust a solution as possible on a tight budget, while investigating various construction methods and materials that would be able to withstand the extreme weather conditions over the course of the year. Because their hometown is located close to a container terminal, the two were inspired to use old shipping containers. This meant that the house would have all the basic functions of any other home, but in a minimal form: a kitchen, dining room, living room, washing and clothes area, and two beds. Stankey and Nordby transported a pair of 20-foot containers to the site and installed them on individual foundations to serve as the sides of the house. Between them, they placed a timber construction with generous steel-framed windows in order to provide sufficient lighting and support the projecting roof structure. The used shipping containers were hardly altered from their original form: apertures cut into the side walls link the containers to the central structure to form a common space, while narrow but strategically positioned window slits offer views of the surrounding landscape.

05 06 07

Project
SAUNA BOX
Architect
CASTOR DESIGN
BRIAN RICHER AND KEI NG

Project location   Various
Estimated use   Sauna
Container type   Freight container

01

02

03

01 – 03
This movable sauna-in-a-box can be situated anywhere: a backyard, a backstreet, or even a snowy clearing.
04
The existing container doors swing open to reveal a smaller, more traditional door leading into the sauna.

04

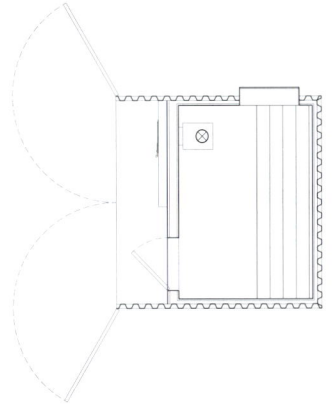

This mobile, stand-alone architecture holds a traditional, wood-fired sauna that was built into a 10-foot steel container. The COR-TEN steel container has standard dimensions and components so that the sauna can be transported to any location without requiring a specially constructed site with specific utilities. An integrated wood oven, rainwater storage and rooftop solar panels make the unit self-sufficient. Aesthetically, it is marked by the interplay of its variety of materials: on the exterior, Castor clad the box with robustly weather and seawater-resistant steel; inside, the sauna is lined with warm wood that corresponds to the overall function of a sauna. The variety of materials is also emphasized by the furnishings that were selected, including stone furniture and wash basins, wooden and metal fittings and fixtures. When the sauna is in operation, the container doors swing open, enlarging the diminutive entrance where the oven opening, a shower and seating are located. The sauna cabin itself, is finished with wood as is typical and has a window opening that, visible on the facade of the container, becomes clear evidence that this structure is not being used in the predictable manner. The architects refuse to mass-produce the Sauna Box, offering instead to fabricate each one to meet the specific requirements of each client, adding bespoke sound systems, guitar amplifiers, lighting or furnishings. The interior design is tailored to suit the advantages associated with the reuse of a freight container, in no way compromising the mobility that is the container's virtue.

Project
FUTURE SHACK
Architect
SEAN GODSELL

Project location   Various
Estimated use   Emergency housing
Container type   Freight container

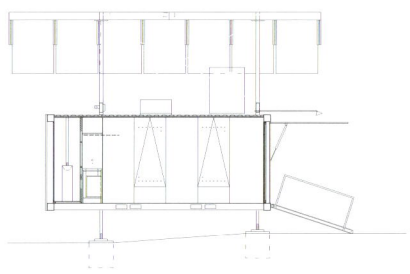

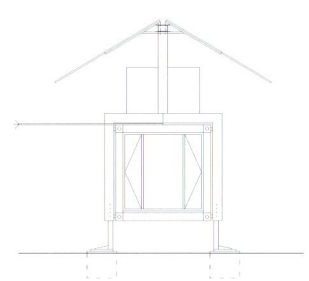

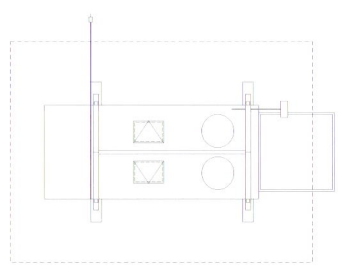

04

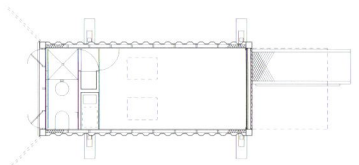

05

Looking like one of NASA's old Apollo lunar landers, Future Shack is actually a mass-produced relocatable house that can be assembled within 24 hours for emergency and relief purposes. Godsell recycled shipping containers to form the main volume of the building. The shack's parasol roof packs inside the container for shipping and when erected shades the container, reducing the building's heat load. The roof panels can also be interchanged with indigenous materials such as thatch, mud and stick, palm leaves and so on. Legs telescope from the container enabling it to be situated easily and without excavation on uneven terrain. The house has a variety of applications following flood, fire, earthquake or similar natural disasters, or functioning as temporary or relief lodgings. Future Shack offers extreme flexibility since as a universal base module, containers can be stockpiled for use on an as-required basis and then transported by trucks, ships and trains worldwide. Easy to assemble on difficult sites, the shack comes packed with a pair of steel brackets, containing four legs that are fixed to the outside of the container. The units are admirably self-sufficient, arriving with water tanks, solar power cell, satellite receiver, roof access ladder, container access ramp and parasol roof, all stored inside the container during shipping. The basic container is also modified to provide thermal insulation and to allow the free flow of fresh air through a series of vents.

01
The Shack ships to its site packed with its components, including a parasol roof that reduces its heat load.
02, 03
Interior with kitchen and bathroom door panels open (top) and closed (bottom) with beds that fold down from the walls.
04
Longitudinal and cross section.
05
Roof elevation and floor plan.
06 [page 103]
Steel brackets and feet make the building easy to assemble, even on difficult terrain.

Project
INFINISKI MANIFESTO HOUSE
Architect
JAMES & MAU ARCHITECTS

Project location   Curacavi, Chile
Estimated use   Private residence
Container type   Freight container

01

02

03

01 — 03
For Infiniski James&Mau repurposed mobile pallets to dress their highly graphical Manifesto House.

Infiniski, a green architecture and design company based in Spain and Chile have built a reputation on houses constructed from recycled shipping containers, pallets and train rails, even going so far as to design a large Chilean office building using salvaged shipping containers. James & Mau Architects' 160-square-meter Manifesto House for Infiniski is intended to represent the brand's concept, as well as its potential in terms of bioclimatic design, recycled and reused materials, nonpolluting construction systems, and the well-considered integration of renewable energy. Rescued maritime containers form the basis of the project, which relies on bioclimatic architecture to adapt the form and positioning of the house to its energy consumption and is based on a prefabricated, modular design that makes its construction both cheaper and faster. James & Mau divided the house into two levels, cutting a single container in half. The first story is used to support the container on the second level. This bridge-like approach to the structure generated a void between the staggered containers, which the architects isolated with thermo glass panels in order to provide a natural ventilation system. Also, as if it had a second skin, the house "dresses and undresses itself", the architects explain, depending on its need for solar heating, thanks to ventilated external solar panels that clad the walls and roof. The house's "skin" consists of sustainably harvested wooden panels on one side and recycled mobile pallets on the other. The pallets can open themselves in winter, allowing the sun to warm the metal surfaces of the containers, and close themselves in summer to protect the house from overheating. Both the exterior and the interior of Manifesto use up to 85% recycled, reused, and eco-friendly materials, including recycled cellulose and cork for insulation, recycled aluminum, iron, and wood and ecological paints.

04, 05
Interior views with exterior wood panels open, leaving large windows unshaded.
06
Wooden panels camouflage rescued maritime containers in the rear while pallets clad the front of the house.
07
The residence uses up to 85% recycled and green materials inside and out.

- 107 -

Project
WIJN OF WATER
Architect
BIJVOET ARCHITECTUUR & STADSONTWERP

Project location   Rotterdam, the Netherlands
Estimated use   Restaurant
Container type   Freight container

01

02

03

04
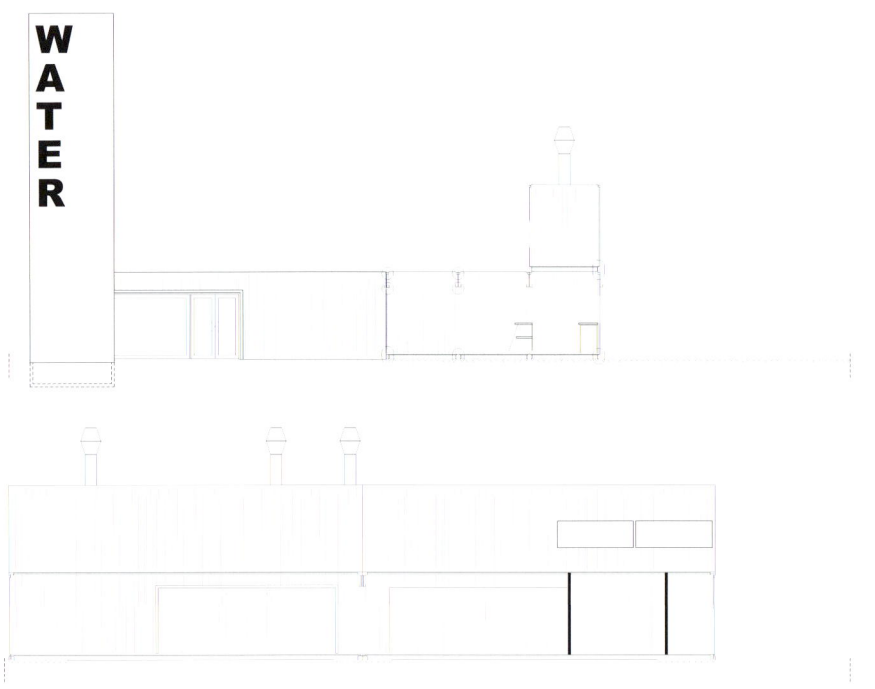

01
Bijvoet salvaged nine freight containers from the nearby port to make this Mediterranean restaurant distinct.
02 – 04
Eight containers were lain horizontally; one, stacked vertically, becomes a tower that advertises the 50-seat venue and forms a small courtyard.
05, 06
Container doors become window shades and openings for generous views of the Maas River and port.

Dutch architect Caroline Bijvoet designed the temporary Wijn of Water restaurant in Rotterdam port. The exclusive Mediterranean cuisine and seating for 50 guests was targeted at the multicultural, creative and highly educated residents of the adjacent neighborhood. Views of the port and Maas River, along with a small budget and short construction timeline pointed towards using maritime elements in the architecture. Bijvoet used nine second-hand, 40-foot freight containers, eight of which she painted in a uniform light blue and stacked in two stories, arranged in an L-shape to form a southeast-facing atrium that is enclosed on three sides. The ninth container stands vertically, advertising the eatery. The position and shape of the building emphasizes the contrast between the built-up city to the north and the open, industrial views to the south, while providing protection against the prevailing westerly winds. To make the best use of the confined height of the containers, all equipment was installed in metal tubes on top of the roof of the building and full surface glazing was applied at the ends of the containers to allow light inside.

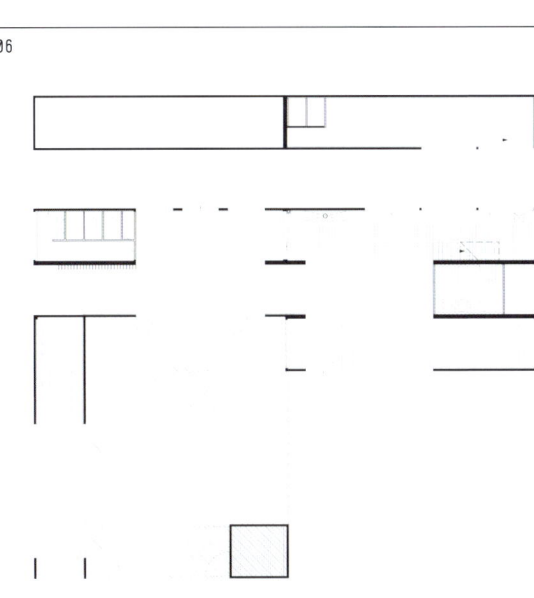

Project
CHILDREN'S ACTIVITY CENTRE
Architect
PHOOEY ARCHITECTS

Project location   Melbourne, Australia
Estimated use   Adventure playground
Container type   Freight container

01, 03
Construction waste generated by slicing the recycled containers was reused for awnings and balustrades.
02
A tiled corridor. Phooey recycled 90% of its materials from site demolition and local sources.
04, 05
Containers were arranged to create both public and more intimate play areas, as well as to connect visually and physically with the surrounding landscape, a potential playground in itself.

04

05

The Children's Activity Centre is located at Skinner's Playground, behind a housing block in South Melbourne. Using the playful name "Phooey Architects", the theme of play is celebrated in this space through function, materiality and aesthetics. The brief was to produce a low-budget recreation centre with after-school facilities that could stand up to hyperactive kids—and to consult the community during the design process by initiating a local workshop around the project. The children themselves requested a two-story cub house with distinct spaces for boys and girls. In addition to these suggestions, Phooey decided to aim for zero waste, reusing containers that they assembled in a staggered arrangement that established both intimate and public spaces suitable for study, painting, dancing, and lounging. Each container was oriented strategically to produce visual and physical connections to surrounding playground areas. The architects turned the construction waste produced by slicing containers into a design feature, using the material for structural balustrade, awnings, and decorative cladding. This architectural approach translates the sustainable concept of feeding back into the building's lifecycle. Cannibalised and repositioned super-scale graphics give new life and character to the shipping containers while a timber deck, staircase, overhangs, awnings and bulk insulation shade the building and reduce energy use. Phooey also salvaged 90% of the materials used—including doors, windows, joinery, carpet tiles, timber, steel, and hardware—from site demolition, local council facilities and recyclers.

06

07

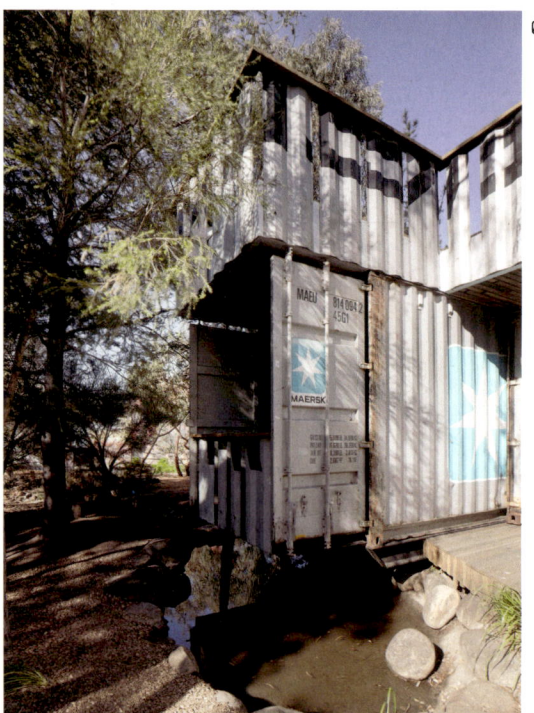

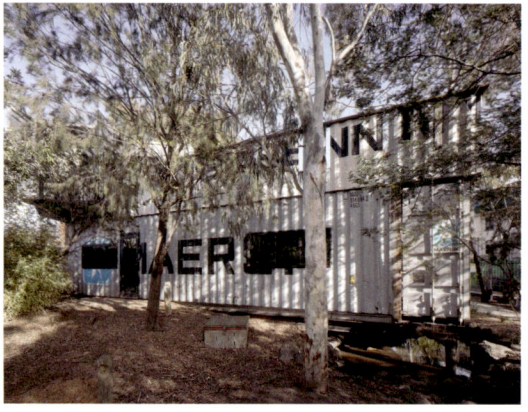

06
Both floors overhang a small stream on the site.
07, 09
Pirated and repositioned type and graphics add another layer of whimsy to the recreation centre.
08
A cross section of the two-story building.

08

09

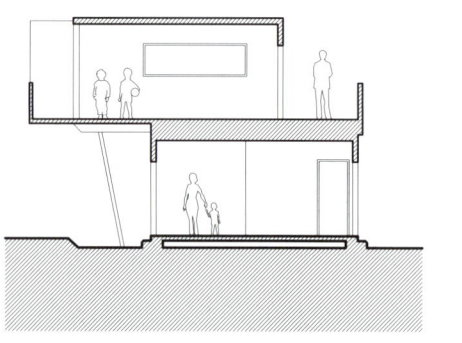

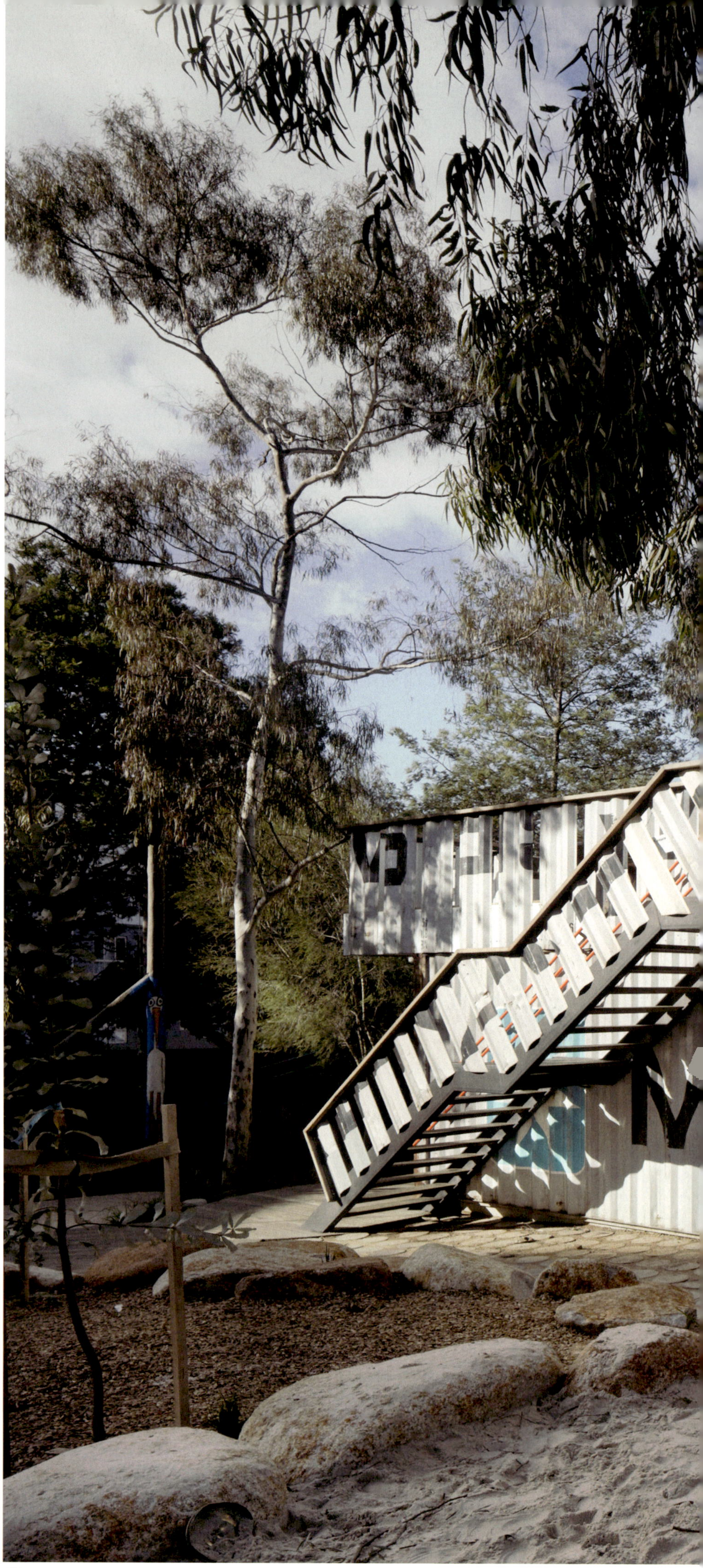

Project
TEMPORARY EXHIBITION / 100 YEARS OF FOOTBALL CLUB ST. PAULI

Architect
KOMMA4 ARCHITEKTEN, PÜTZ / REETZ

Project location   Hamburg, Germany
Estimated use   Exhibition space
Container type   Freight container

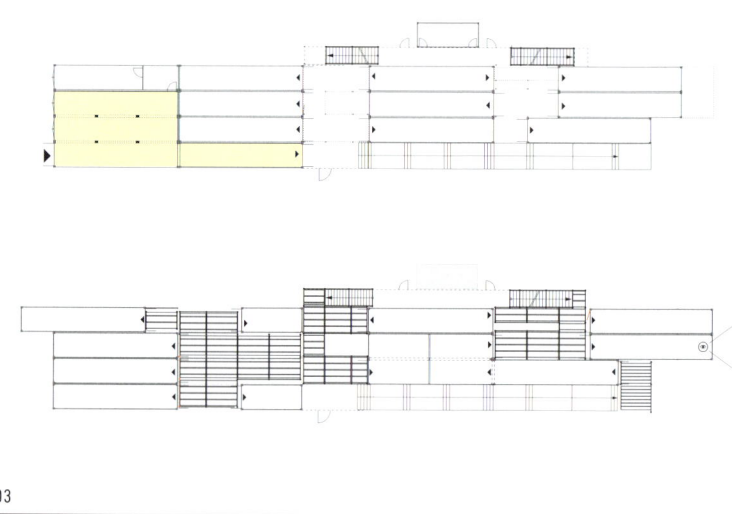

Proving that Hamburg's port-side St. Pauli quarter isn't just about hookers, hemp and hooliganism, Komma4 will erect a temporary exhibition building at Hamburg's cultish (if modestly successful) football club St. Pauli. The exhibition that it houses will present the left-leaning history of the local Millerntor Stadium in the form of publications, films, photographs, documents and objects that have long been spread across the globe. The core element of the new building will be an assembly of 40 sea containers that will be temporarily turned into "see containers" where the artifacts will be presented over 500 square meters of floor area, spread over a number of floors. The architects settled on container architecture not merely to define the indoor space, but also to open it up to the outside as a stage for events that will take place over the course of the event.

01, 02
Renderings of the building in the urban context.
03
Floor plans.

Project
INTERSTITIAL LIVING
Architect
LHVH ARCHITEKTEN

Project location   Cologne, Germany
Estimated use   Private residence
Container type   Building container

01 – 03
A solution to the lack of affordable housing, these prefabricated modular living units can be plugged into city services on a variety of sites, from disused lots to rooftops.

A lack of affordable real estate close to the city led LHVH to try a new approach to living and working in the city's "in-between" spaces. The architects put unused infill areas, empty lots, roofs, and raw building containers to use at central locations, occupying them with prefabricated, inexpensive modules. Existing containers were particularly well-suited for this purpose and allowed easy access to utilities due to the modular nature of the container and the existing supply network, which ensured that basic energy sources were accessible. The project was first presented to the public during the plan05 architecture festival under the rubric of "Living In The In-Between", documenting various potential uses for and configurations of these infill shelters over a period of one year.

Project
PLATOON CULTURAL DEVELOPMENT BERLIN
Architect
PLATOON.BERLIN & SOEREN ROEHRS

Project location   Berlin, Germany
Estimated use   Communication platform and office space
Container type   Freight container

01 – 03
The creative agency paired stacked freight containers with a matching patch of lawn and lap pool around which to host office parties and cultural events.
04
Site plan.

Founded in 2000, the Berlin offices of this multidisciplinary creative organization—whose projects focus on reality rather than the typical consumer wish to escape from it—are made, aptly enough, from freight containers. The 110-square-meter "communication platform" and office is made from four freight containers with the addition of a suitably railroad-track-shaped swimming pool. The containers are framed with plywood inside and planked with aluminum. One thing is made clear by this real-world industrial presentation of the company headquarters: "PLATOON does not entertain. PLATOON communicates." They just have a pool deck to do it on.

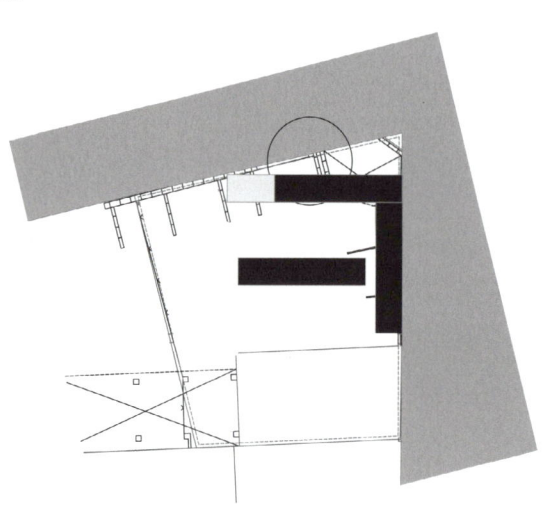

05

05
Platoon's offices are one of their best bits of branding yet.
06
The site of the Platoon headquarters in Berlin.
07
Capping the ends of the containers with glass fills each floor with natural light.

Project
CONTAINER HOUSE (KILLER HOUSE)
Architect
ROSS STEVENS

Project location   Wellington, New Zealand
Estimated use   Private residence
Container type   Freight container

01
Stevens built this three-story part-time residence from pre-insulated containers and tower crane sections beside the local dump.
02
The bedroom.
03
Salvaged fire escape stairs and scrap steel fixtures and finishes.
04
Detail of the spiral staircase.
05
Clad in grey, the house exploits its site's geological flaws.
06, 07
The building leans into the embrace of three rock walls.

Stevens just wanted "a small basic place to sleep" when he had to work in the city, away from his rural home. Pursuing an ecologically and economically sustainable design while responding to the geological flaws of the site, inspired the architect to re-use refrigerated (pre-insulated) containers, tower crane sections, fire escape stairs, and pieces of industrial waste steel, all neatly assembled on the site of an abandoned hole in an excavated rock near the local rubbish dump. The building was designed to fit snugly into a space between three rock walls that give the dwelling shelter from wind and security when it is unoccupied. This idea led Stevens to limit the number of windows on the front of the building; instead he opted for 20 optical lenses that allow views out but not back in. Once inside, the rock becomes part of the building interior, allowing the house to feel both subterranean and open. Because the rock is not entirely stable, occasionally falling onto veranda floor, the designer forged the floor from 50-millimeter-thick industrial waste steel selected precisely to endure the impact. Incredibly, the site is also located on an active earthquake fault line that is expected to cause serious damage to the city in the next 50 years (a fact of which the fractured and falling rock serves as a constant visual reminder) further justifying Stevens' use of industrial products in the construction of a strong yet lightweight shelter.

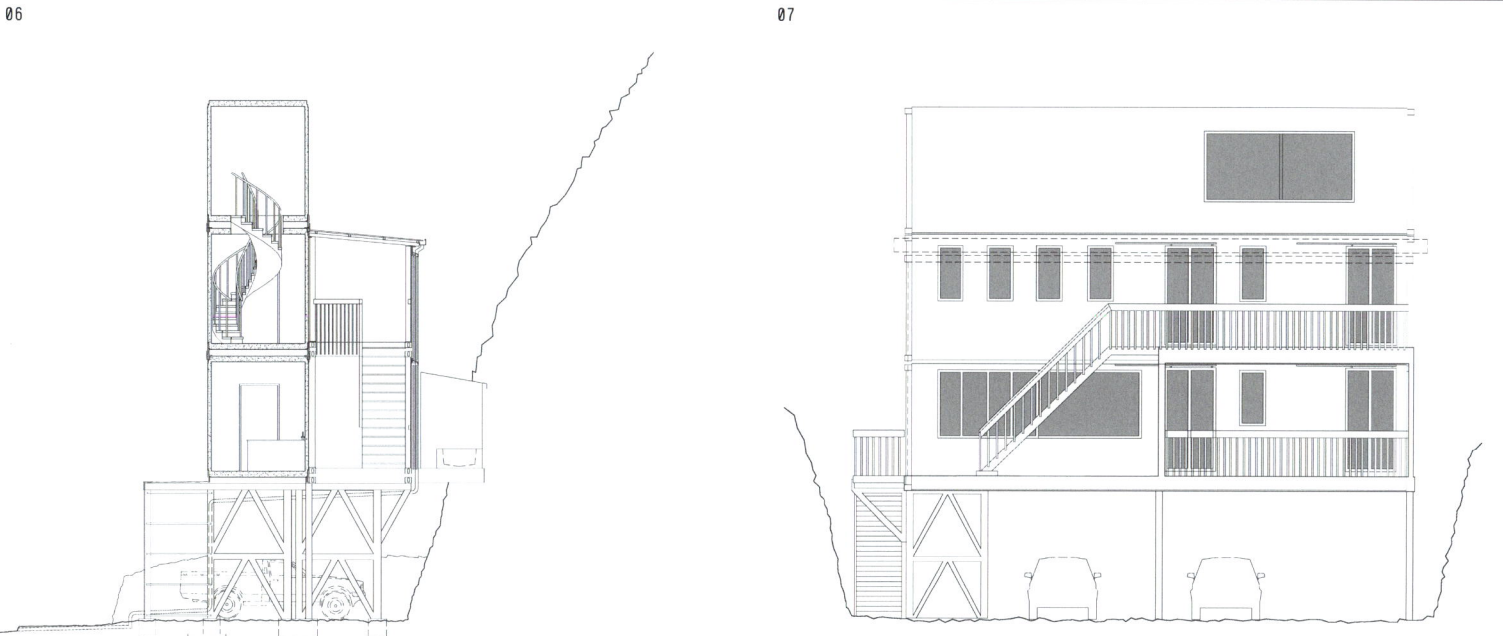

06

07

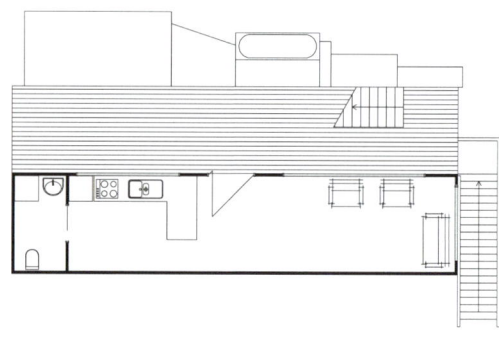

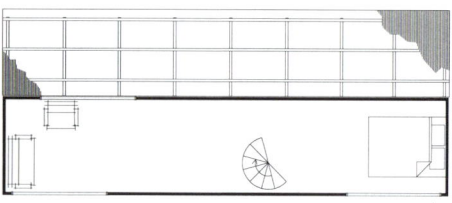

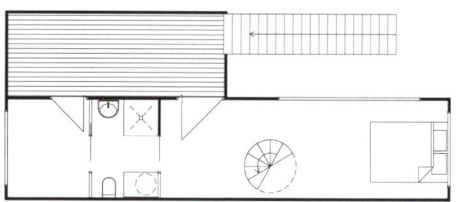

08, 09
Cargo numbers and caution signs allude to the former functions of Stevens' construction components.
10, 11
Floor plans.
12
Scrap metal and the living rock wall that buttresses the house are decorative elements in the design.
13
The industrial salvaging means that the interior sometimes recalls an old-fashioned train carriage.

12

13

Project
FREITAG FLAGSHIP STORE ZURICH

Architect
SPILLMANN ECHSLE ARCHITEKTEN

Project location Zurich, Switzerland
Estimated use   Retail space and observation platform
Container type   Freight container

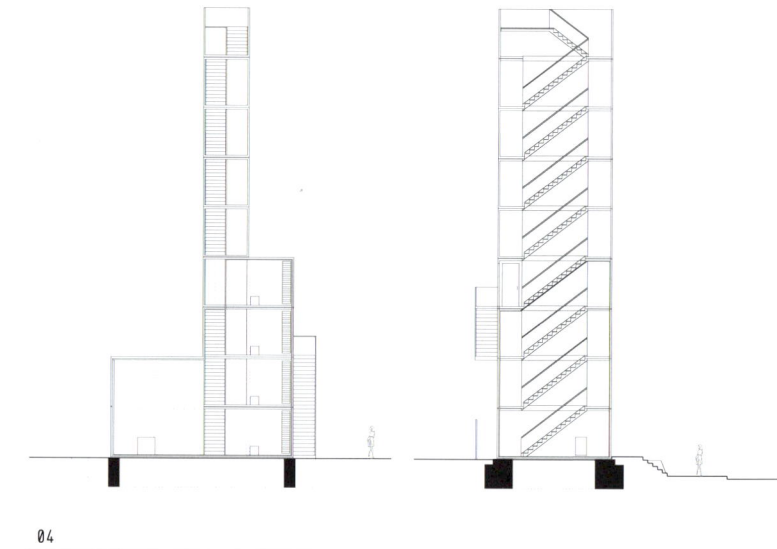

01
Located on a regional transport hub, Freitag's Zurich flagship is a tower of freight containers that echoes the brand's ecofriendly ethos.
02, 03
Interior views of retail and showroom space.
04
Spillmann Echsle stacked 17 containers, fixing them in place with shipping industry fasteners.
05 [page 123]
An observation platform tops the showrooms 25 meters above the Geroldstrasse neighborhood.

As much landmark as showroom, the 120-square-meter Zurich Freitag Flagship Store comes crowned—and crowns the city—with an observation platform. The architecture represents an extension of the brand's recycling ethos: since 1993, Freitag has manufactured bags made from repurposed truck tarpaulins, an idea inspired by the company's view of the nearby bridge. Spillmann Echsle architects found 17 freight containers in Hamburg and of course, shipped them to Zurich by rail. They then stacked the colorful boxes (fixing them with shipping industry fasteners) low enough so as to not violate the city's restriction on high-rise construction in the surrounding Geroldstrasse neighborhood, in which the shop is located. The area has long been shaped by its location at the center of several transportation lines (including both rail and auto) and its industrial fabric: a low-slung, smattering of buildings and highly trafficked storage infrastructure. The retail space is situated at the base of the stack, while the tower contains showrooms. Generous window openings saturate the interiors with natural light and allow views in both directions, from the inside out and the outside in. Shoppers circulate past the full product range, which includes more than 150 bags, and then ascend to the observation platform 25 meters above ground with a view of the city, the ribbons of traffic, and Zurich's picturesque alpine lake. The design preserves the same view onto the bridge that gave birth to the Freitag bag concept in the first place.

Project
LIVE/WORK SPACE
Architect
SCULP(IT) ARCHITECTEN

Project location   Antwerp, Belgium
Estimated use   Private residence and office space
Container type   Container look

01
On a tiny 2.4-meter-wide Antwerp lot, the entire storefont of this architecture office swings open to serve as its door.
02
The living room on the second floor.
03
Kitchen and dining room look out from the first floor.
04
Each wooden floor was inserted shelf-life into a frame set between th existing walls of this remnant city parcel and is lit with a unique colour at night.

    In Antwerp's red light district, not all the lights are red. Architects Pieter Peerlings and Silvia Mertens of Sculp(It) live and work in a remnant space that is a mere 2.4 meters wide, 5.5 meters deep and 12 meters high and have created a light installation out of their home. Each of the four wooden floors, which were inserted between two existing walls and slotted into a steel frame, take on their own hue at night. The ground floor serves as the architects' studio and is illuminated in white, the kitchen/dining room on the first floor is green, the living room on the second floor glows red, while the third-floor bedroom is lit blue. The building's shelf-like floors are connected by a spiral staircase that was tucked into the space in a single piece. Three black polyethylene tubes, arranged vertically and horizontally and housed between double wooden floors, provide the infrastructure for plumbing, heating and electricity. Outfitted with a porcelain bathtub, the roof is dedicated to (sun)bathing. The street-level office is fitted with a glassy storefront which is also a door that is as wide as the site. In fact, the designers glazed the whole facade with large, black-framed panes that give a nod (and perhaps a wink, too) to the professionals who formerly occupied this neighborhood—the prostitutes.

Project
MEETINGPOINT PLAN06
Architect
LHVH ARCHITEKTEN

Project location   Cologne, Germany
Estimated use   Exhibition space
Container type   Building container

01, 02
This three-level meeting point for an architectural event used stacked building containers.
03
Cross sections and elevation drawings

This temporary exhibition space for the plan06 architecture festival, was the product of the organizers' wish to locate the event's traditional meeting point, not in a museum, empty department store, or old industrial hall, but in the city's popular Belgian quarter near galleries, shops, restaurants, clubs, courtyards and squares. The event center was built in a three-story container tower on the Brüsseler Platz square, where visible from afar, the intervention sharpened people's perception of their familiar surroundings. It also offered a new perspective of the streets of this Wilhelminian era quarter from its top floor. Furthermore, additional containers were distributed in car parks throughout the neighborhood with the goal of integrating the events into the fabric of this creative urban district. The large display windows cut into the containers were used by participants to exhibit their projects to guests of the fair, as well as to incidental passers-by.

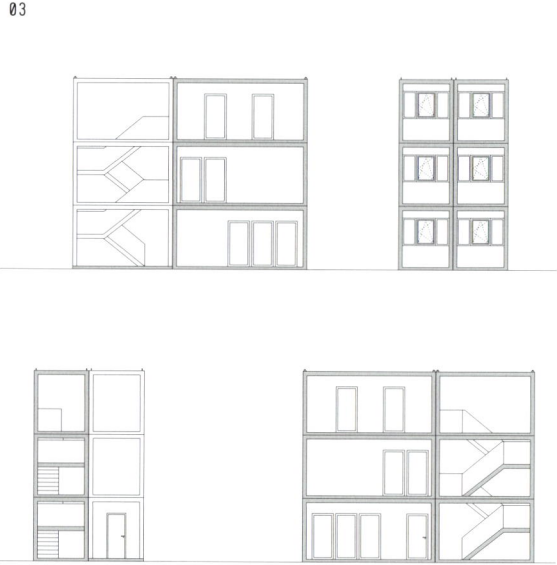

Project
HALF MILE
Architect
LEIBNIZ UNIVERSITY HANOVER
HILDE LÉON AND UDO WEILACHER

Project location   Hanover, Germany
Estimated use   Installation and observation platform
Container type   Freight container

01, 02
This two story installation featured a rooftop deck.
03
Contour lines completed the installation and threaded the installation into the city fabric.

On the occasion of the Night of Sciences as well as the 175th anniversary of the Leibniz University Hanover, the approximately 800 meters (half mile) stretch between the faculties of architecture and landscape architecture was reshaped by a spatial installation related to the cityscape under the direction of Prof. Hilde Léon and Prof. Udo Weilacher. An irregular pattern of white chalk lines was applied to the entire urban space, thus contracting areas allocated in different ways (streets, rails, parking lots, pedestrian pathways, lawns etc.) to form a single terrain. The pattern, a structure comprised of topographical isoheights, so-called isohypses, was perceivable from a normal human perspective only in a fragmentary way—and therefore at first, inaccessible to viewers. In order to grasp the installation in its entirety, heightened vistas were needed. These were provided in the form of three temporary viewing platforms made of the simplest materials and built without being treated in any way: stacked freight containers, simple scaffold staircases, and materials required for scaffolding such as railings for platforms.

Project
CONTAINR CINEMA
Architect
ROBERT DUKE ARCHITECT
WITH KIETH DOYLE, IAIN SINCLAIR, NICOLE MION AND EVANN SIEBENS

Project location	Vancouver, British Colombia, Canada
Estimated use	Event space
Container type	Freight container

01, 02
A pop-up theatre and sustainable architecture.
03
A number of Vancouver artists contributed to the development of this cultural container-in-a-container.
04
The boxes were stacked, staggered and painted.
05
Inside, surfaces were cut away to create a double-height ceiling.

In Vancouver's Public Library Square, a consortium of local artists under the direction of architect Robert Duke, developed containR as a temporary way to bring art and ideas directly into the public realm. The project combined video, public art and sustainable urban design. As a branded object, it is a modular portable event space that can be dropped into place and assembled within several hours. containR is a pair of unevenly stacked freight containers that have been cut away to create a double-height interior, with a projector and screen tucked into opposite ends of the vaulted void of the upper container. The upper container is cantilevered beyond the lower one by 10 feet, defining a dramatic exit and entrance while solar panels have been integrated into the architecture, doubling as an aesthetic element. From the outside, passers by during the city's 2009 Cultural Olympiad viewed graphic painting and watched a video monitor that screened a documentary about the object's construction. In the future, the multipurpose box will also be programmed for use as a gallery, retail outlet, DJ/VJ club, live stage and possibly even low-cost housing, says Duke.

05

Project
EICHBAUMOPER
Architect
RAUMLABORBERLIN
JAN LIESEGANG AND MATTHIAS RICK

Project location   Mülheim an der Ruhr, Germany
Estimated use   Event space
Container type   Freight container

01

02

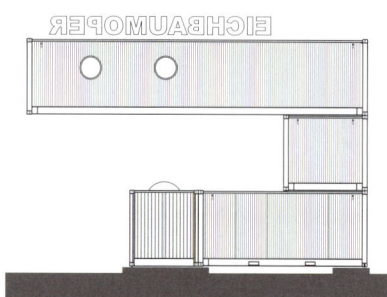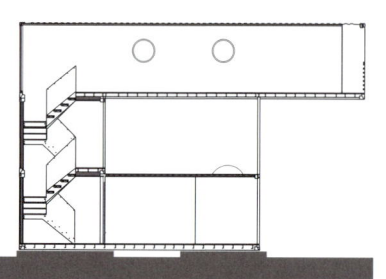

Through four humble shipping containers, raumlaborberlin architects Jan Liesegang and Matthias Rick transformed a problematic underground station that had long suffered from vandalism and violent assaults into a 50-square-meter opera house. The project represented a communal effort to rehabilitate a beleaguered public space through architecture, art and theater. The three-story space contained an office and workshop, stage, a room for community meetings and a "laboratory." With Liesegang and Rick as artistic directors, the container frame system, with partial steel-structure reinforcements, stood for one year. Inside, a new type of opera was created in an unusual collaboration between designers, musicians, artists and local people. Part of the libretto was taken directly from resident's stories of waiting at the station and when the performances began in mid-2009, 60 residents participated as extras. Normal train service was maintained during show times, and some even took place inside the trains en route between Essen and the Hirschlandplatz stop in Mülheim.

03

04

05

06

07

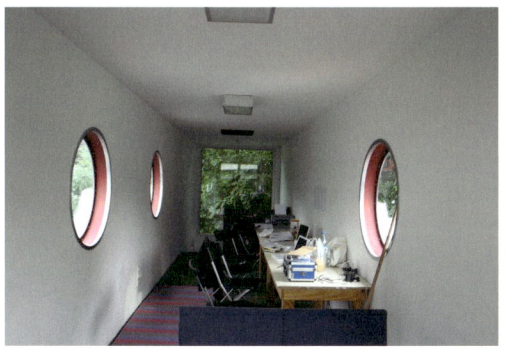

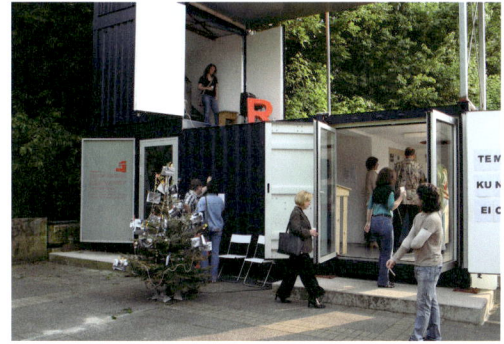

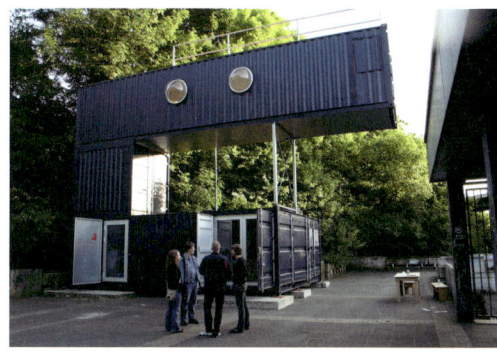

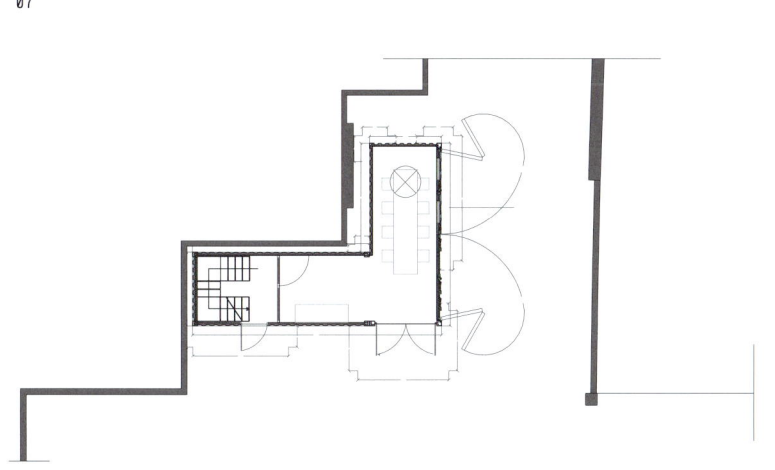

01
The container-based cultural space gave new life to a crime-ridden rail station.
02
The project seen in section.
03
The space closed up.
04
Interior office space.
05
The space open to the public.
06
Exterior view.
07
Site plan showing the unusual arrangement of volumes.

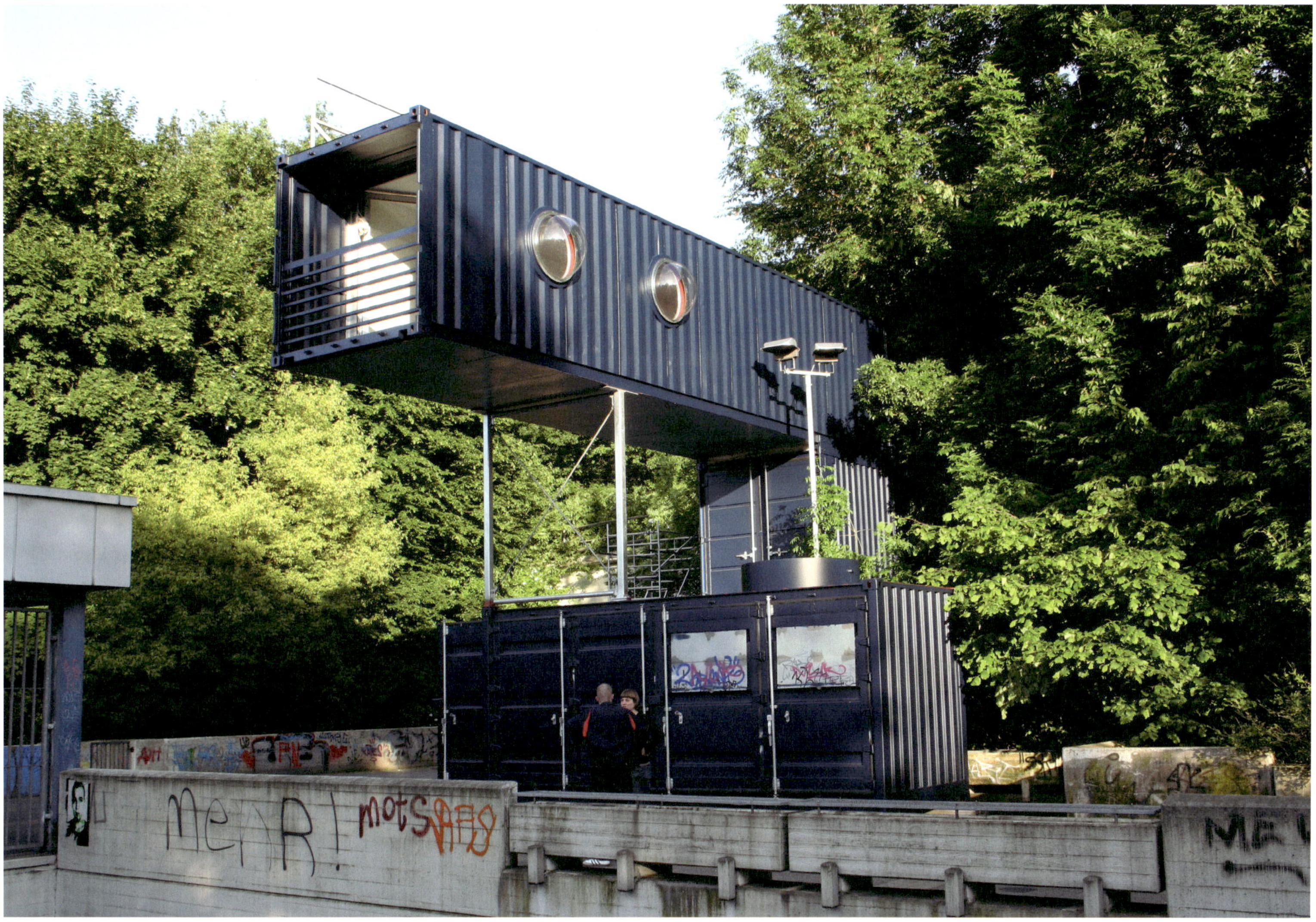

Project
ORBINO
Architect
LUC DELEU

Project location   Amsterdam-West, Noordzeekanaal, Netherlands
Estimated use   Sculpture and observation platform
Container type   Freight container

01

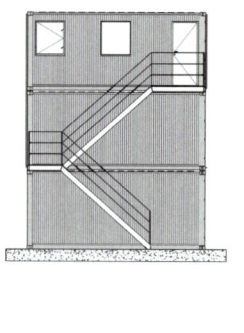 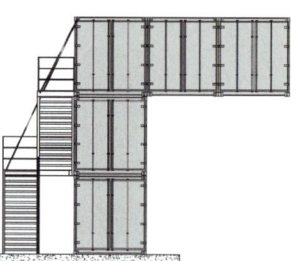

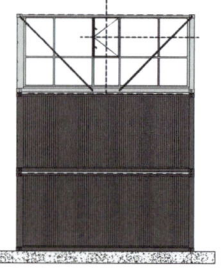 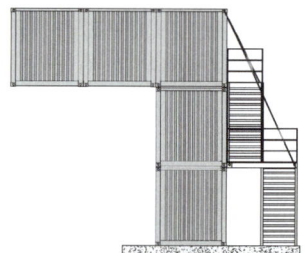

02

Orbino is a permanent, 45-square-meter open-air sculpture on the Noordzeekanaal in Amsterdam. Belgian architect and artist Luc Deleu's walk-through art work is made from three rehabilitated shipping containers with screw bridge fittings, stacking cones, a steel staircase, and a concrete base. It plays with the relationship between horizontal and vertical. Deleu arranged the containers in a precarious pile: the L-shaped tower resembles a periscope because the upper containers cantilever out over empty air by two container-widths. A steel staircase mounted to the exterior leads to a room formed by the three highest containers, two-thirds of which consists of the cantilevered box. With floor-to-ceiling glazing on one side, this interior acts as an enclosed observation platform.

01, 03, 04
A periscope-like cantilevering of containers serves as artwork and vista in Antwerp.
02
Orbino seen in elevation drawings.
05
The observation platform is accessed via a staircase anchored to the rear of the tower.

03

04

05

Project
CONSTRUCTION X, HOORN BRIDGE, SPEYBANK
Architect
LUC DELEU

Project location	Antwerp, Belgium (Construction X and Speybank)
	Hoorn, the Netherlands (Hoorn Bridge)
Estimated use	Scuptures and bridge
Container type	Freight container

01
Construction X.
02
The Hoorn Bridge.
03
Speybank.

Over the past two decades, Belgian artist Luc Deleu has created sculptural and architecturally oriented pieces by integrating industrial containers into his work in forms that range from A to Z. Made for the 1990 exhibition For Real Now, Construction X on the Hoorn Bridge in the Netherlands worked triple time as an artwork and a 30-square-meter pedestrian and bicycle bridge. Deleu removed the rear walls of two 20-foot containers and, as he is wont to do, used screw bridge fittings and special interconnecting devices to fix the containers in place. The Speybank sculpture resembled a bridge, with four impossibly balanced containers forming a vast horseshoe that seems ready to collapse at any moment. Construction X, on the other hand, was a less functional, albeit more stable looking, construction.

Project
CONSUMER TEMPLE – BROKEN ICON
Architect
GARY DEIRMENDJIAN

Project location   Sydney, NSW, Australia
Estimated use   Sculpture
Container type   Freight container

01
The view upward into the artwork.
02
Looking down to the pallet floor and "buttresses".
03
Deirmendjian cut openings into one wall of the container to create a church-like dimness.

The work sits in the green expanse, a singular presence in a state of stable decay. A narrow entrance leads into an intimately walled enclosure which offers no way out other than that of entry. Only fragments of the outer world are discernible through the walls. In the relative darkness, the eye is drawn up to the airy chamber high above, where light beams through small windows onto a centrally suspended, broken cash register. The container is held upright with its opening facing down, atop four stacks of timber pallets. A combination of ground anchors, cables and tensioning secure the structure to the ground. An enclosure is created by walling off the three sides of the four pallet stacks, using a variety of pallets and commercial crates—a combination that also allows for seating. Small rectangular apertures cut into the container's timber floor, allow limited light in.

Project
CONTAINERART SÃO PAULO
Architect
ARTUR LESCHER, BERNARDES+JACÓBSEN

Project location	São Paulo, Brazil
Estimated use	Exhibition space
Container type	Freight container

ContainerArt promotes exhibitions of contemporary art that is displayed inside disused shipping containers. Sometimes the containers are scattered around a city center, along its beach, outskirts and cultural locations. At other times, they are clustered together to create temporary art hubs in a museum setting, as in the case of the Vila Lobos project. This 25 × 33 meter version of the ContainerArt concept, the so-called "hub-spoke" model by Brazilian artist Artur Lescher, with local architects Bernardes+Jacobsen, for the client Automatica, borrows the best practices from intermodal logistics and applies them to enhance the visual and display impact and networking potential of an art event. In São Paulo, ContainerArt served as a temporary museum hub dedicated to video art for ten days. Because of its utilitarian but elegant composition, which provides wheelchair accessibility along with protection from both sun and rain, variants of this structural approach were later adopted by subsequent ContainerArt projects.

01 – 04
Stacked two high and offset to form an arcade, these containers serve as temporary art space and can be configured to suit the site and the content of the exhibition.

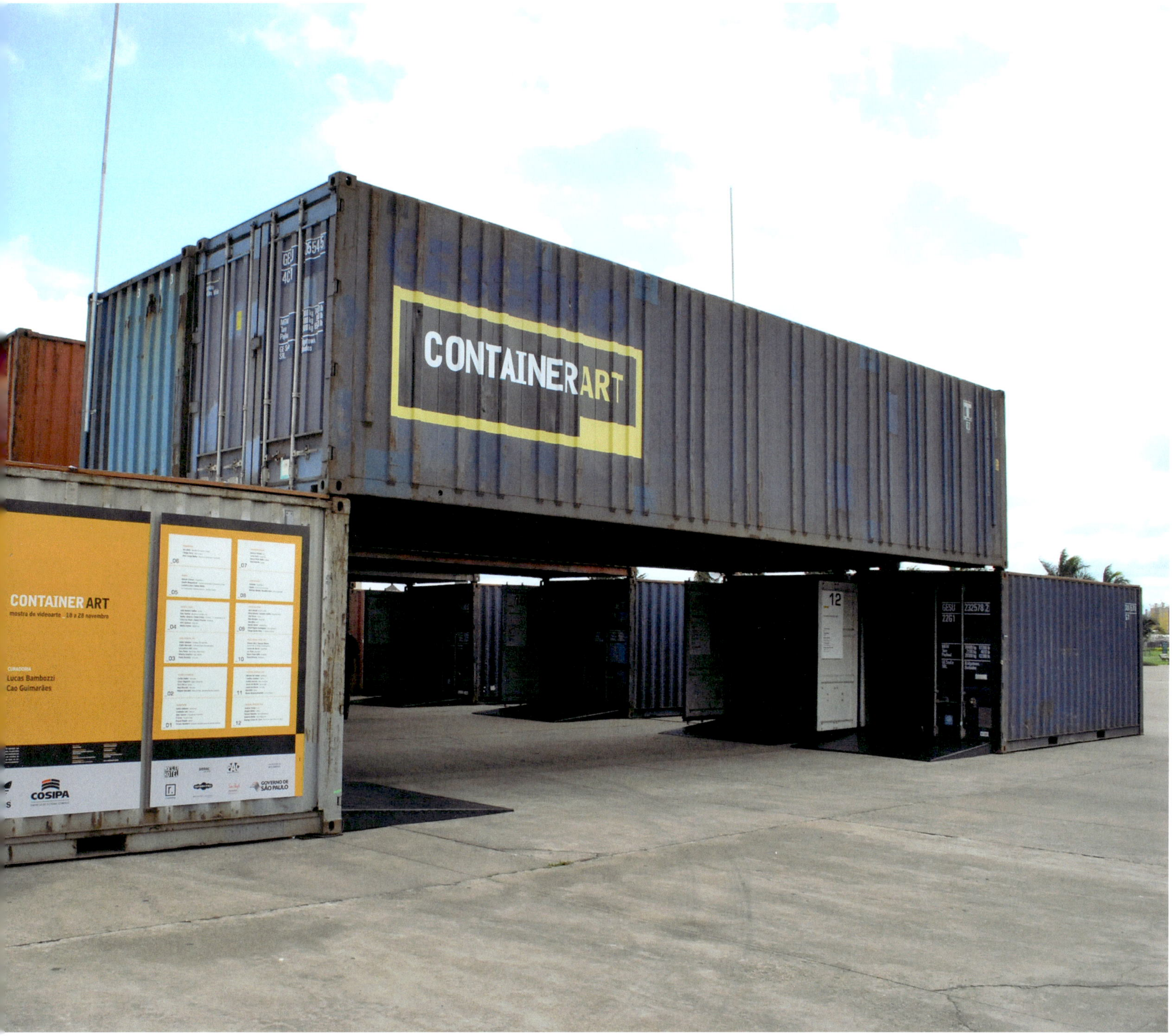
CONTAINERART
CONTAINER ART
mostra de videoarte _18 a 28 novembro
CURADORIA
Lucas Bambozzi
Cao Guimarães
COSIPA
GOVERNO DE SÃO PAULO

04

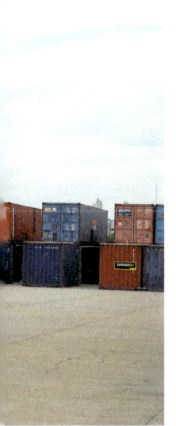

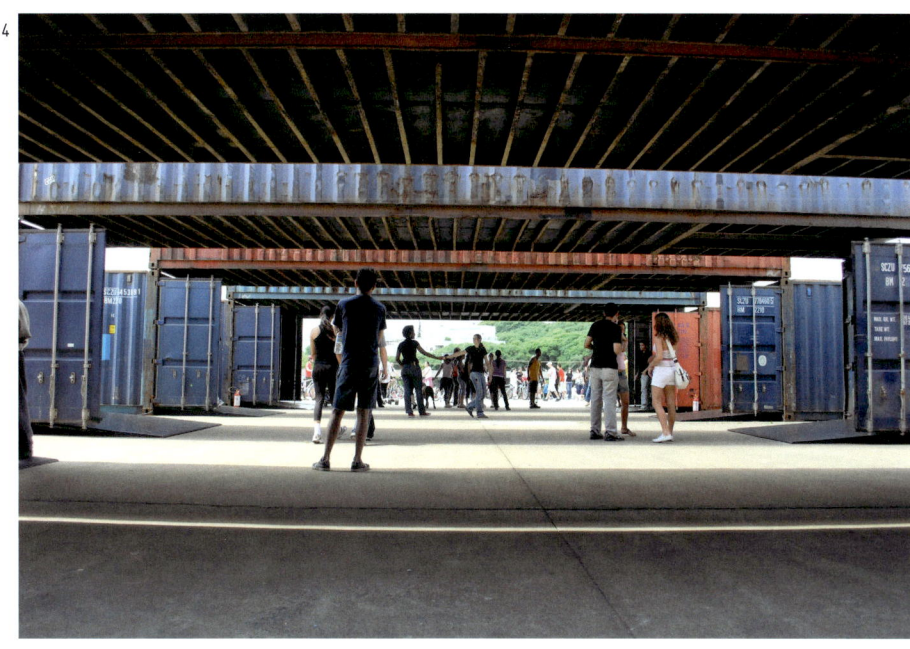

Project
GAD
Architect
MMW ARCHITECTS OF NORWAY

Project location   Oslo, Norway
Estimated use   Exhibition space
Container type   Freight container

This "semi-temporary" gallery can be disassembled, moved and reassembled easily at any location and within just a few days. MMW drafted seven ordinary steel containers into service, building the ground floor from two of them, surrounding the central first floor courtyard with three others, and using the final two to give access to the top-floor balcony. Industrial ladders and stairs connect the containers, since they are a part of the main circulation system throughout the space. Because its components overlap slightly, the composition of the building became light and airy, while bestowing on GAD a great deal of protected and unprotected outdoor space. The interiors of the containers are insulated and covered with sheets of plywood and sheetrock, all painted white to give the gallery clean display surfaces. The containers are opened up via circular windows placed opposite each other—skylights and floor-to-ceiling safety glass windows at the ends of each container—letting large amounts of northern light into the building. With its extroverted personality, GAD lends itself admirably to an invigorating display of art.

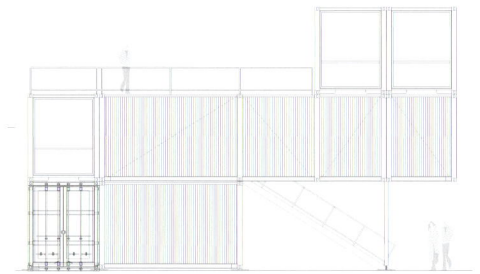

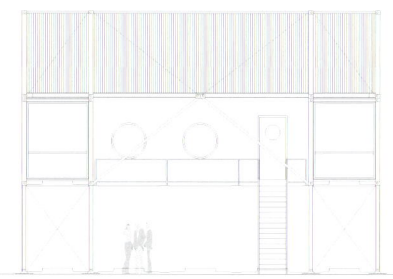

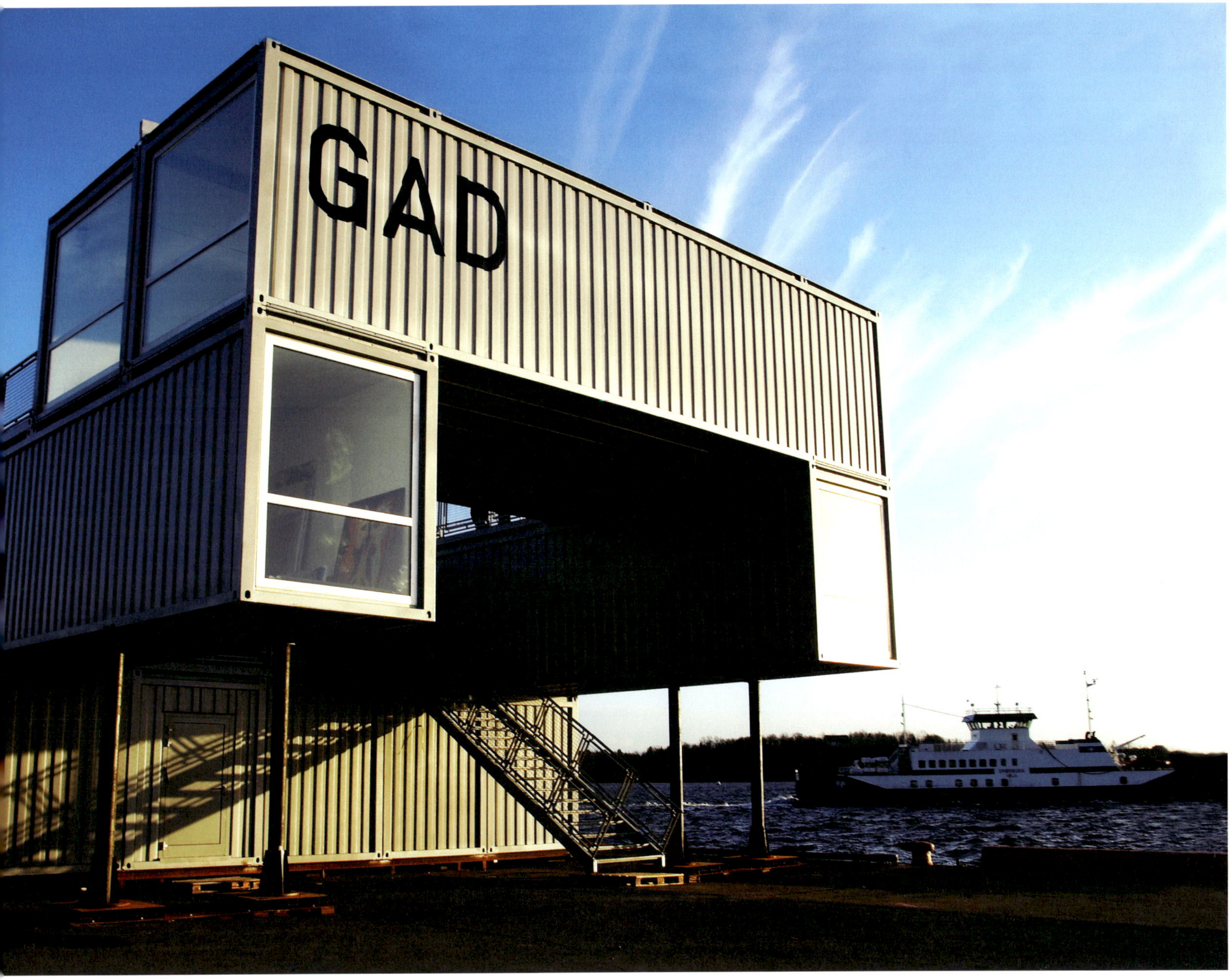

01
A rendering shows the circular skylights on the gallery's ground and first floors.
02, 05
The overhanging containers give the building a dynamic tension.
03
The view through a first floor gallery.
04
Apertures cut into the containers form doors, skylights and nice views.
06 – 08
Floor plans.

06

07

08

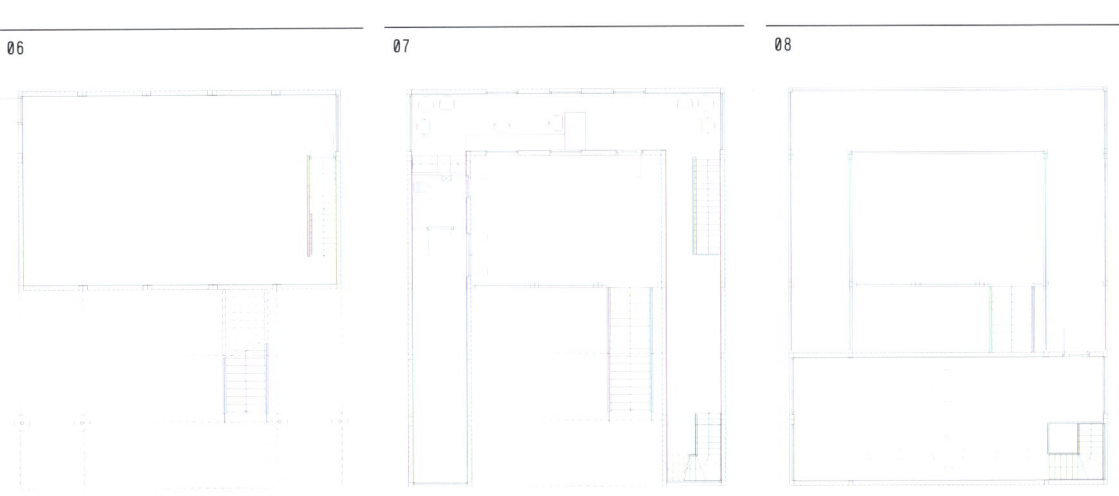

Project
PUMA CITY
Architect
LOT-EK

Project location   Alicante, Spain
                   Boston, Massachusetts, USA
Estimated use   Retail and event space
Container type   Freight container

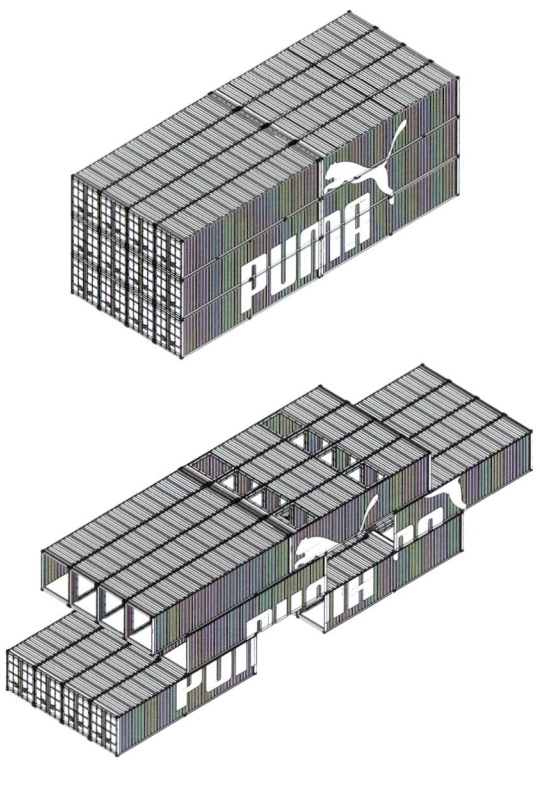

01, 03
The fragmented logo, pulled apart by shifting the stack of containers, advertised the architecture as much as Puma City, itself.
02
Fully glazed ends of the containers let in light and framed views to the interior.
03
The stack shift created double-height ceilings on the lower floors and open spaces that let in light, air and circulation routes.
04
Two retail spaces filled the lower levels with offices and storage on the first floor and a terrace and event space at the top.
05, 06
Section drawings.

Taking full advantage of the global shipping network already in place, New York studio LOT-EK retrofitted 24 cargo containers to transform them into Puma City. Puma City was a movable 1,020-square-meter retail and event space that, along with the 70-foot long Puma sailing boat, il mostro, toured the world during a year-long marketing campaign for the athletic gear retailer. Like the sailboat, the City was fully demountable and traveled on a cargo ship. Ada Tolla and Giovanni Lignano conceived of the building as a three-story stack of containers, staggered to create internal outdoor spaces, large overhangs and terraces. The stack was then branded with the super-graphic logo of the company which appeared intriguingly fragmented as a result of the stack shift. Puma City featured two full retail spaces on the lower levels, both with large double-height ceilings, as well as with four-container-wide open spaces to challenge the modular boxy quality of the containers' interiors. The second level was occupied by offices, a press area and storage while on the top floor, a bar, lounge and event space opened out onto a large terrace. The construction used 40-foot shipping containers as well as a number of the existing container connectors to join and secure the units both horizontally and vertically. Each module was designed to ship as a conventional cargo container through a system of structural covering panels that fully sealed all of its large openings. These seals were then removed on site to reconnect the large, open interiors. With this project, LOT-EK challenged the scale at which a building could be truly mobile, and designed Puma City to confront all of the architectural obstacles faced by a building of its kind, including international building codes, dramatic climate changes, plug-in electrical and HVAC systems and ease of constant assembly, disassembly and operations.

04

05

06

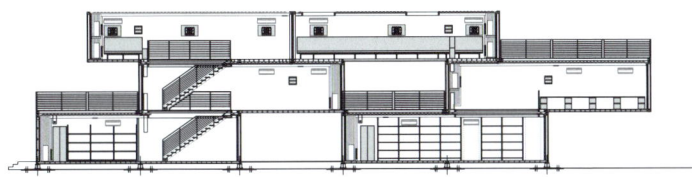

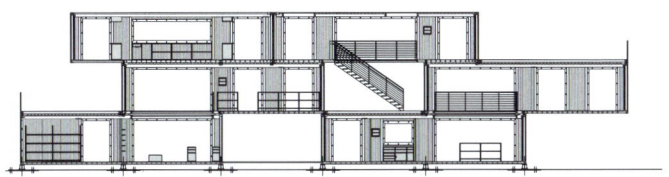

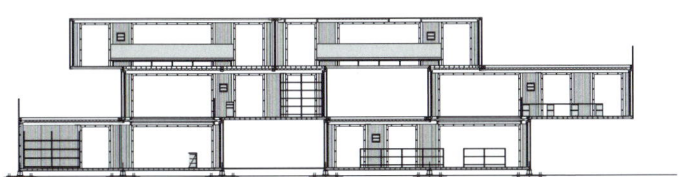

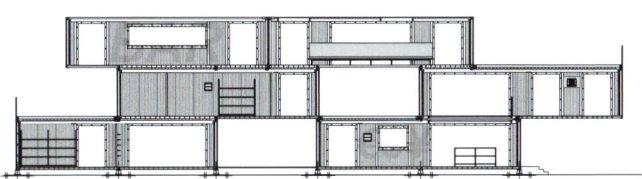

07

08

09

10

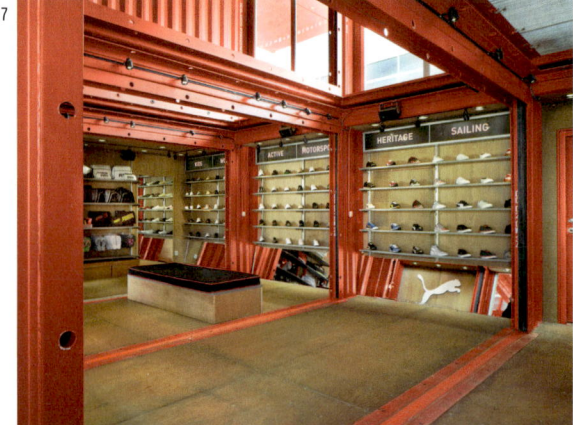

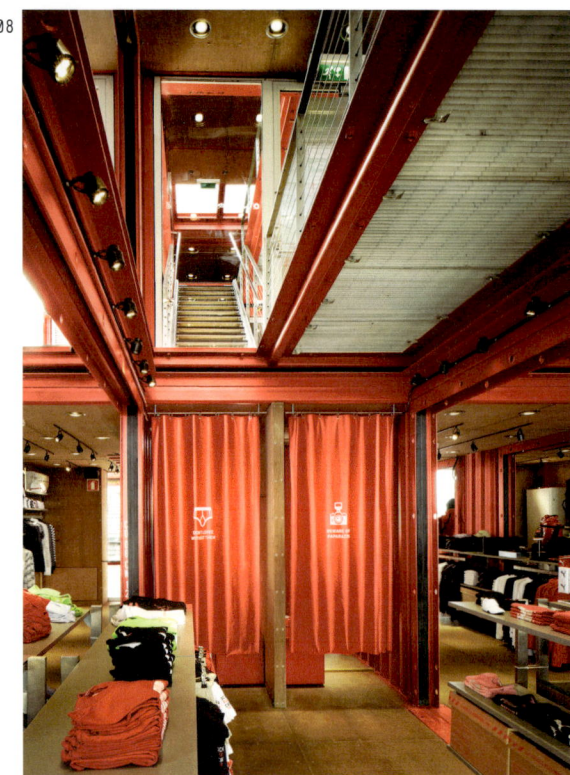

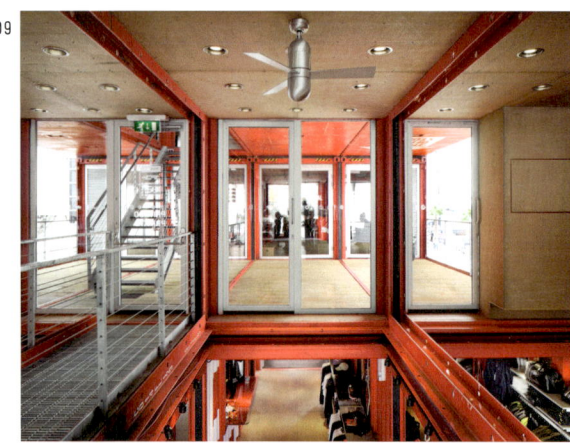

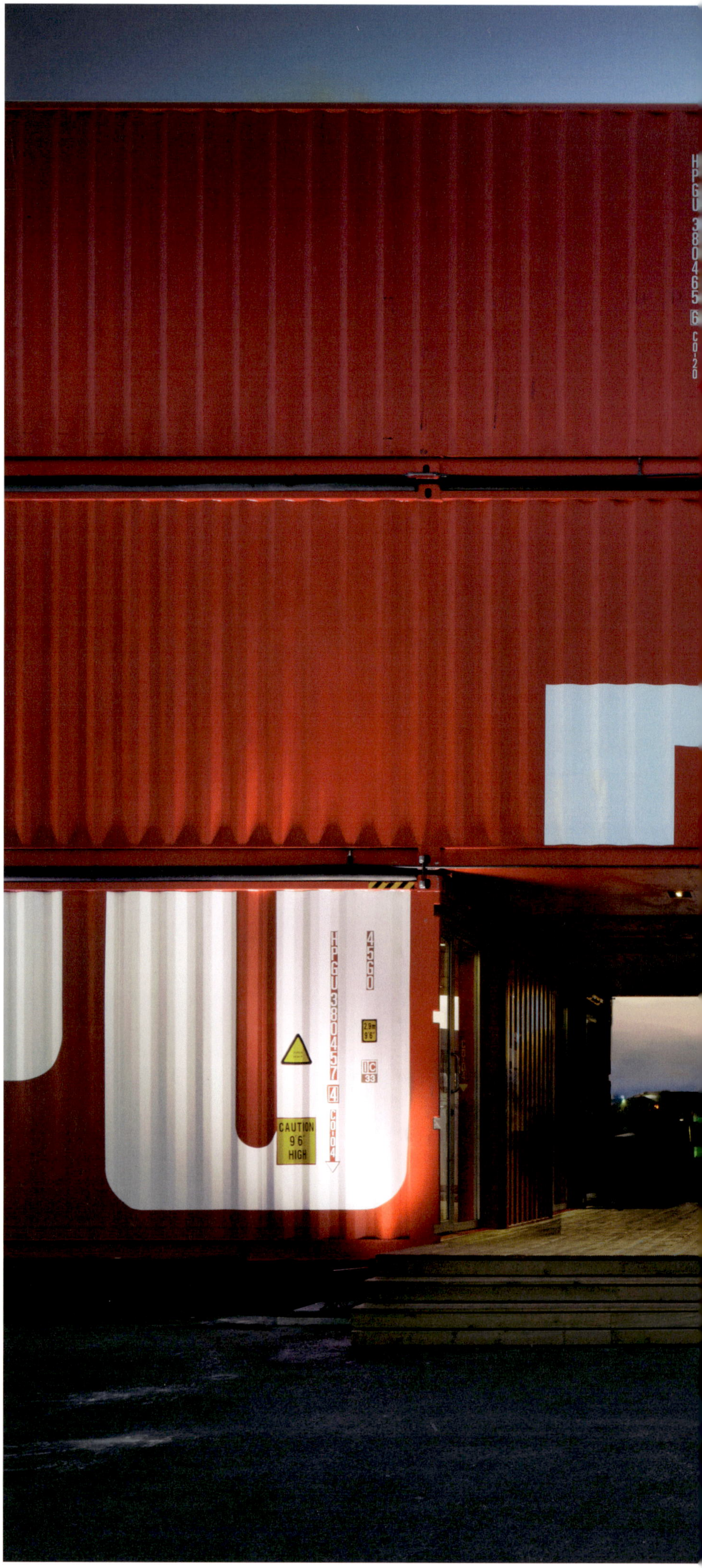

07, 08
Shop interiors on the lower floors.
09
By removing container walls, LOT-EK created high ceilings and greater transparency.
10
An entry arcade created at ground level by the first floor containers.

Project
SUBURBAN HOUSE KIT
Architect
ADAM KALKIN

Project location   New York City, New York, USA
Estimated use   Installation
Container type   Freight container

01
Scenography for the Deitch Gallery exhibition re-created a fantasy suburbia in which a residence comes ready-made in a freight container.
02, 04
Interiors that articulated the wastefulness of suburban living alongside responsible design.
03
The container house kit wouldn't have been complete without the car parked in the driveway.

Dipping into your midlife crisis? It may be time to acquire a new life-in-a-box. The Suburban House Kit is a suburban idyll conceived by Adam Kalkin with artists and designers including Tobias Rehberger and shown at Manhattan's Deitch Projects gallery as an artwork. In addition to a fully equipped and furnished home assembled from freight containers, the house came with an origami garden, a courtyard, a car and garage, and other amenities typical of the suburban American way of life. Best of all, it can be ordered as a complete architectural and lifestyle "kit".

The use of freight containers in this cheeky project—which was simultaneously critical and optimistic—set its ironic tone since it juxtaposed unconventional building materials with the supremely conventional design features of suburbia; wastefulness with responsible living.

Project
OLD LADY HOUSE
Architect
ADAM KALKIN

Project location   Califon, New Jersey, USA
Estimated use   Private residence
Container type   Freight container

01

New Jersey-based architect Adam Kalkin created this 93-square-meter house from freight containers and a large, wrap-around deck. Its simple and energy-efficient, single-story plan was designed to enable an elderly person to be able to live alone and enjoy an easily maintained domestic space. The central box features automatic glass doors at either end leading to the deck from the front and rear of the home and contains the dining and living zones. To one side of the main living room, another container holds the kitchen, laundry room, and a walk-in closet. Flanking the opposite side of the living room are two bedrooms and a bathroom. In some sections, the architect removed container walls, floors or ceilings to create double-height spaces. Due to its intelligent use of space-saving container modules, the house offers a number of spacious interior rooms that are easily accessible and maintainable.

01
This energy-efficient home can be easily maintained by its elderly owner, prolonging her independence into old age.

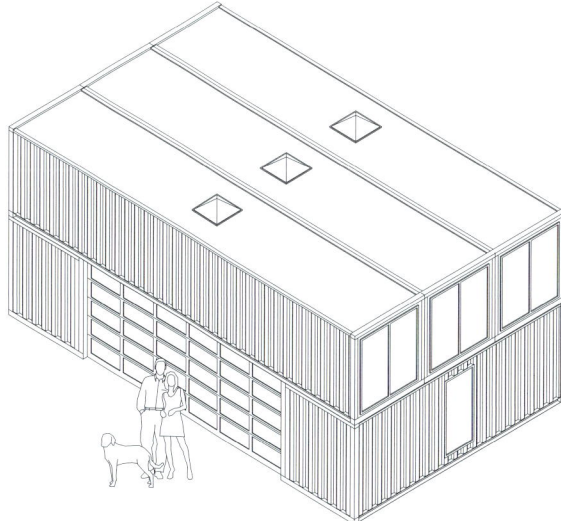

02
A wraparound porch gives easy access to rooms up the slope.
03, 04
Interiors feature generous glazing.
05
Two volumes flank a cosy patio.
06
The patio connects ground floor entrances.
07
The basic unit of the design.

Project
12 CONTAINER HOUSE
Architect
ADAM KALKIN

Project location — Brooklin, Maine, USA
Estimated use — Private residence
Container type — Freight container

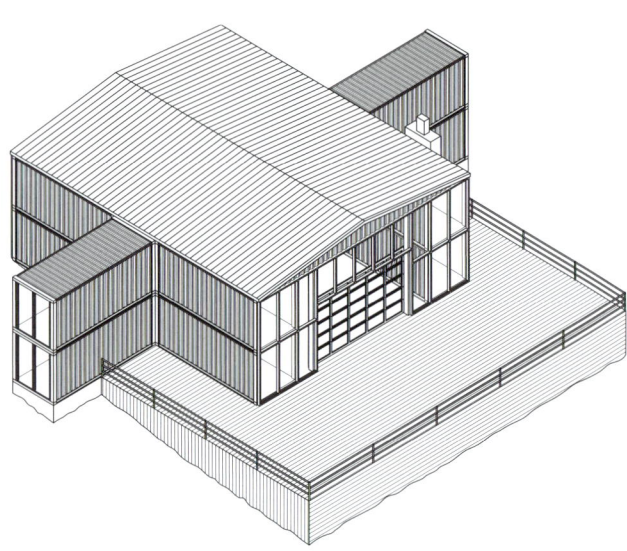

01

04

The 12 Container House, used as a summer and vacation house, was built on the rugged Atlantic coastline of Brooklin in Maine, outside the city and surrounded by trees. Its basic structure follows the principle of combining containers with other structural elements, in this case a prefabricated steel-glass construction forming the central living hall. The house consists of 12 salvaged freight containers that set on a common reinforced concrete slab, form the exterior walls of the central, two-story, roofed living space. They are grouped around the hall to form two T-shaped units. Two containers are stacked upon each other. The ground-floor containers along with the hall accommodate a spacious living, dining and kitchen space, while the upper containers serve as private bedrooms. Open stairs lead from the central living space to the upper story. The front walls of the containers are fully glazed, also toward the interior space, thus offering additional lighting from the hall. The long sides are all closed to the outside, but some are fully open to the inside to integrate the kitchen and dining room into the living room. The gabled sides of the living room can be opened by means of lowering industrial garage doors, thus making it a sort of buffer or intermediate space between the outdoors and the completely enclosed containers.

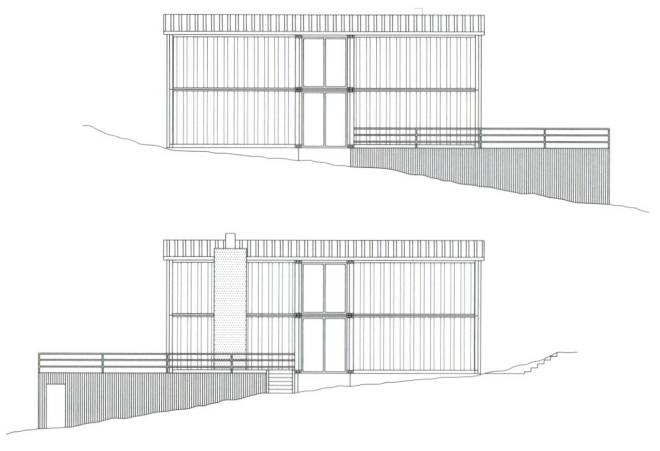

05

06

07

01
An isometric view of the house.
02
Front facade.
03
Rear view with some container walls replaced by windows.
04
The north and south elevations.
05
Looking through the open industrial gate to the dining room.
06
Kitchen.
07
Stair detail.

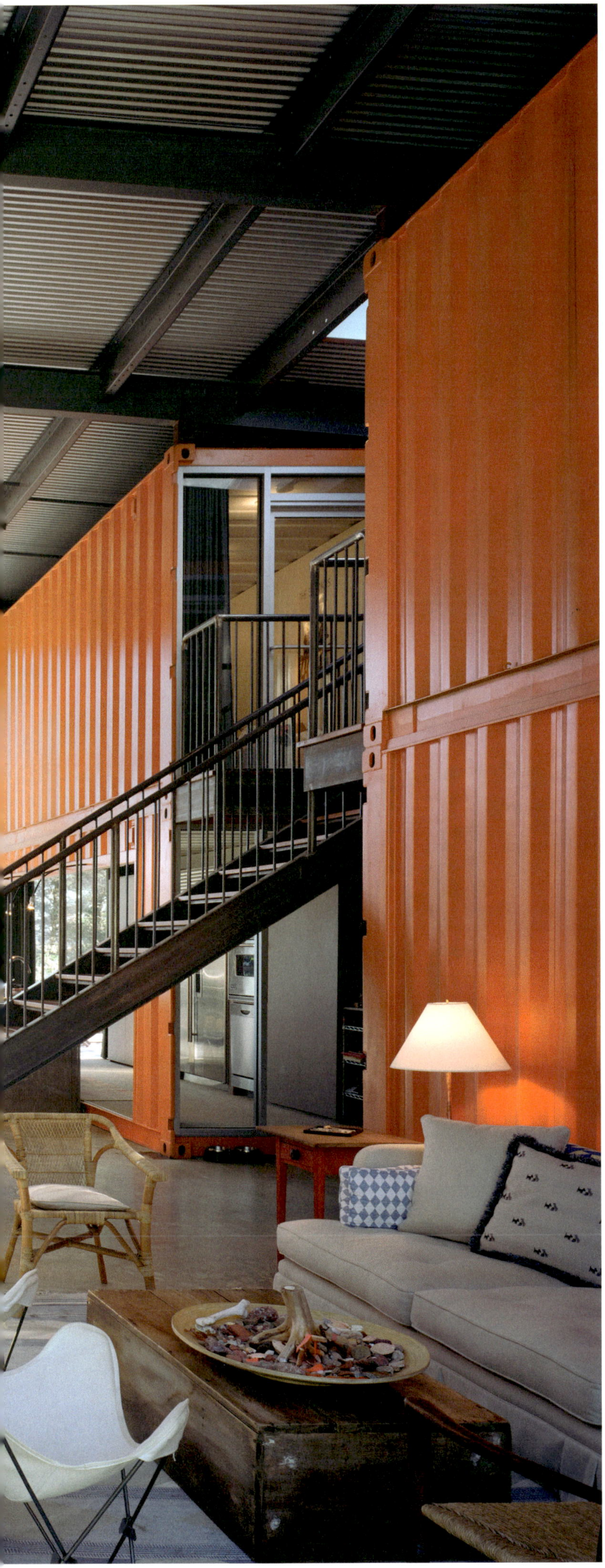

09

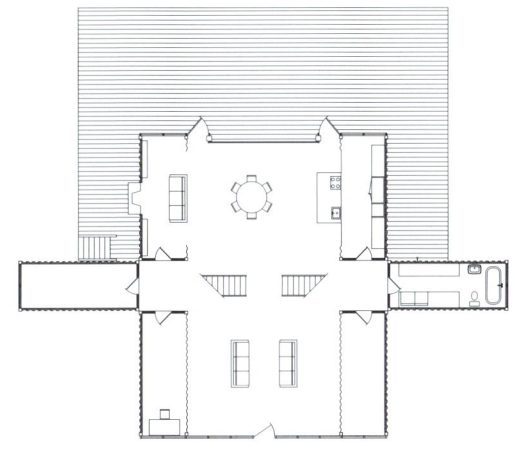

10

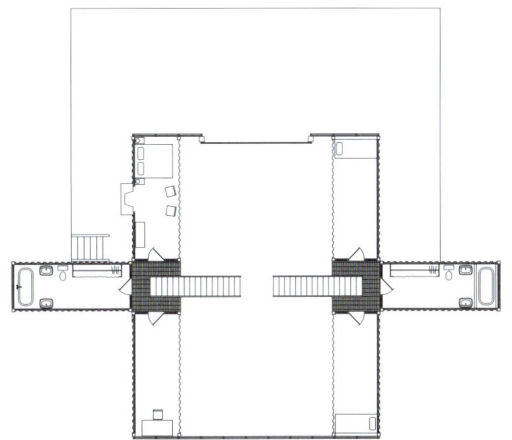

11

08
Twin staircases seem to wink
at the traditional grand stair
in this industrial, modern-day
"mansion."
09
The library niche.
10
Plan of the first floor.
11
Second floor plan.

Project
KING KAMEHAMEHA BEACH CLUB
Architect
ANGELA FRITSCH ARCHITEKTEN

Project location   Offenbach am Main, Germany
Estimated use   Event space
Container type   Freight container

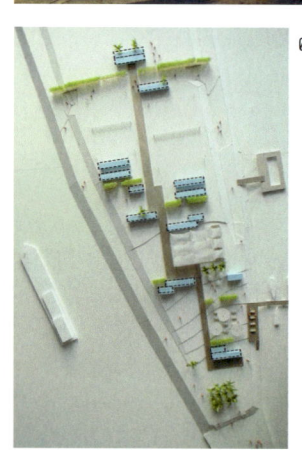

The King Kamehameha Beach Club reclines on a peninsula in Offenbach Port, an offshoot of Frankfurt's Kingkameha discotheque. During the day, the club is frequented by families who come for recreation, while in the evening it becomes an open-air disco. Surrounded by water, the site is parceled into sections dedicated to sports, relaxation, socializing, and so forth through the simple arrangement of used and converted freight containers. Although the main leisure areas consist of open spaces between containers, the containers themselves house the functions necessary for running the facility. At the same time, the containers create a visual link to the club's surroundings, which are characterized by the port activity.

01
The Beach Club bar.
02 – 04
Open spaces between containers offer recreational space while services are housed inside.
05
A model of the site on a peninsula in the city's port.

Project
BOOK FAIR PAVILION
Architect
FRADE ARQUITECTOS
JUAN PABLO RODRÍGUEZ FRADE

Project location	Madrid, Spain
Estimated use	Showroom
Container type	Freight container

01 – 03
Frade used color, transparency and a bit of chaos to give depth, texture and energy to the book fair pavilion he made from rescued freight containers.

Commisioned by the Madrid city council to build its Feria del Libro event space, local architect Juan Pablo Rodriguez Frade followed the slogan "Europe Is Built From Books" and constructed the building with maritime containers. The choice to use containers visually emphasized the concept of "building" while generating an easily legible contrast between the uniform interior and the fragmented, modular exterior that served as its jacket. The pavilion was composed from 11 durable, low-cost and super-low maintenance cargo containers of 20 feet each, which were raised off the floor by means of a pine-wood-plank structure. On the outside, Frade painted the containers in bright hues and stacked them in apparent disorder, as if they had been piled there momentarily and were just about to be removed. By contrast inside, visitors were immersed in a wooden micro-environment that matched the scale of the building to its contents and formalized a number of spatial scenarios that the architect related to the outdoors. In this way, the designer authored a clever narrative that suggested the dual notion of "container and contents", a well-considered metaphor in the context of the book fair.

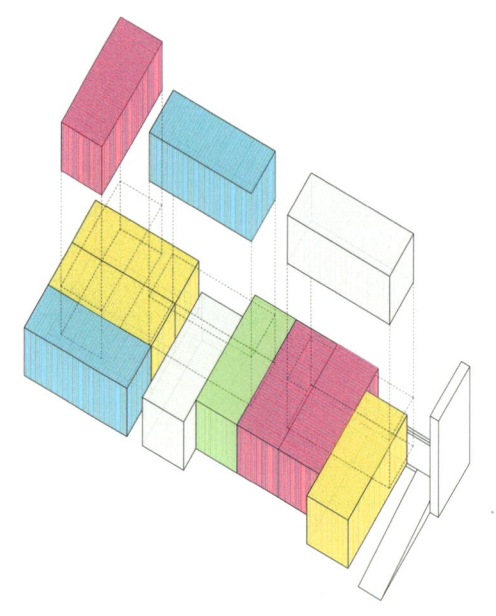

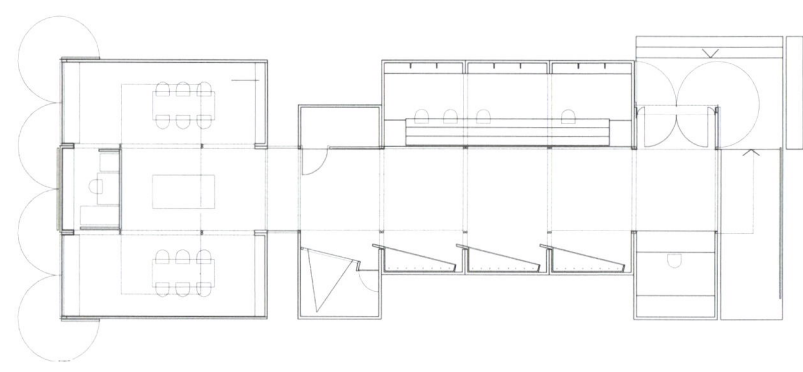

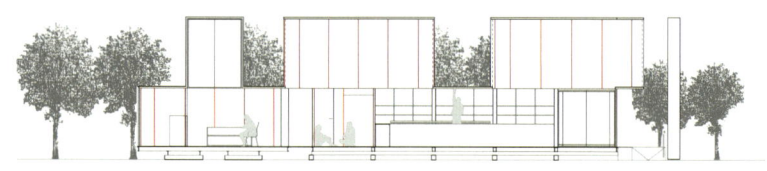

04, 05
The playful exterior never hints at the pavilion's more sober insides.
06
The containers stacked in careful disarray and painted in a vivid colour scheme.
07
Floor plan.
08
View in sections.
09 – 11
The pavilion as a brilliant jumble on the outside in contrast to the more conventional interiors.

Project
B-CAMP
Architect
HELEN & HARD ARCHITEKTEN

Project location   Stavanger, Norway
Estimated use   Private residence and studio space
Container type   Building container

01

02

Norwegian architects Helen & Hard Architekten have worked on several recycling projects which reuse construction materials and components discarded by the oil industry. This experiment has been made easier by the studio's proximity to the site of the largest Norwegian oil reserves, where many international oil companies are located. B-Camp is situated on the coast at Stavanger, on a former industrial site, and involved the transformation of ten discarded industrial containers into four residential apartments, which are intended to serve as a type of "incubator" center for start-up companies and young, creative freelancers. This low-budget dwelling alternative features tiny 25–35-square-meter units, whose diminutive square-footage is balanced with higher-than-usual ceilings, bi-level plans, and individual terraces on the upper levels. The repurposed container units, which were once used as accommodation on their original industrial site, were fitted with a jumble of irregularly framed but generous window openings and additional doors. The units are composed of surplus windows and doors, scrap materials from the metal industry, and clad with a transparent, corrugated plastic panels that have a thermally insulating effect, meaning that only five centimeters of additional insulation was necessary in order to provide satisfactory heat-loss protection. B-Camp offers an alternative lifestyle option for young people who aspire to work in a creative environment and enjoy affordable housing with individual character. In addition, the Helen & Hard offices, which are located close by, also rent out office facilities which include seminar and conference rooms, and social and break areas to B-Camp residents.

01
B-Camp is located on a former industrial site close to Norways largest oil reserves and refineries.
02
The architects cut irregular window openings into industrial building containers to create apartments for young workers.

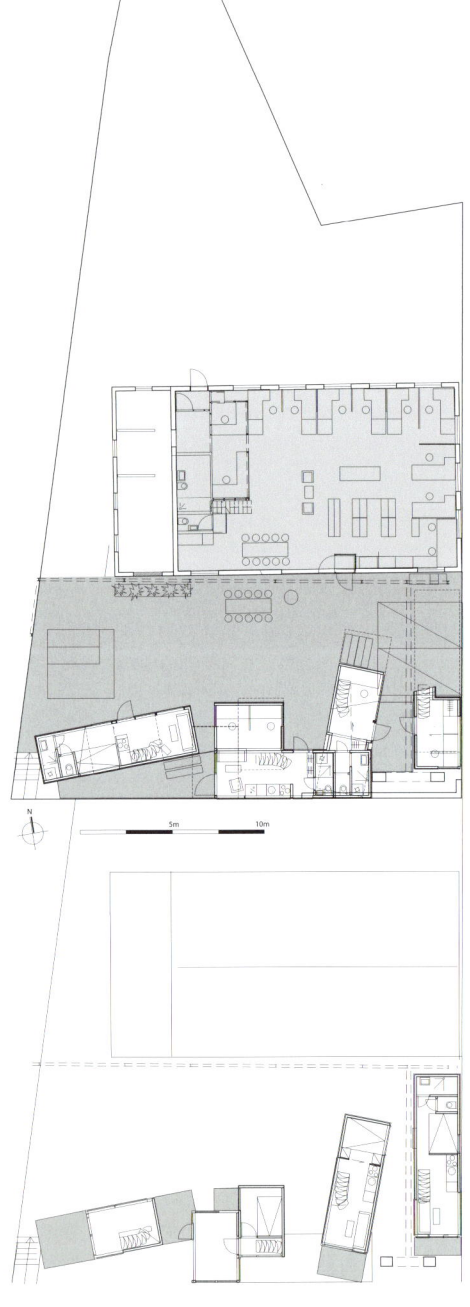

03

05   06

07

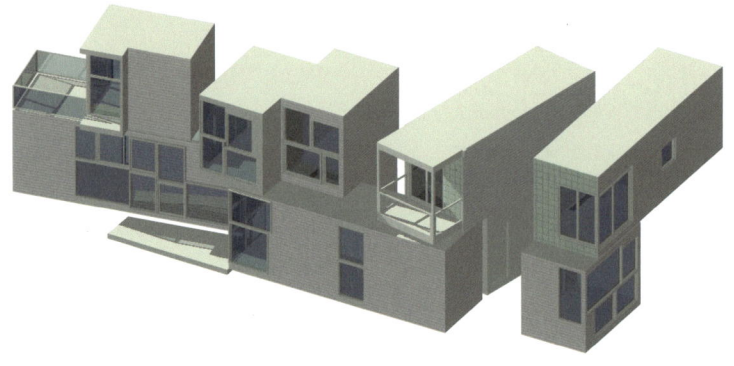

03, 07
Site plans and isometric rendering showing the beautiful disorder of the containers' two stories.
04, 08
Views from outside.
05, 06
Small living spaces are offset with high rooms, bilevel plans and lots of light.

Project
ARCHITEKTURBOX
Architect
SHE-ARCHITEKTEN
ULRICH HAHNEFELD, MARCO PAWLIK, STEPHEN PERRY, STEPHAN SCHRICK AND TORSTEN STERN

Project location Hamburg, Germany
Estimated use   Event space and information center
Container type   Container frame

This information kiosk was designed to stand for six days during Hamburg's Architecture Summer event. In the city center and opposite the city's Gallery of Contemporary Art, the delightfully odd structure, with a cafe that sat astride a small terrace overlooking the water, served as an information center for the public, a promotional hub for events and destinations, and a rest stop for fair visitors. Architects Torsten Stern, Marco Pawlik, Stephen Perry, Stephan Schrick and Ulrich Hahnefeld envisioned the building as a pavilion for visitors to the city and the event, as well as a forum for exhibitions, lectures and discussions. The team decided to pile container frames instead of tidily stacking them, giving the structure textural intrigue against the skyline and city fabric. The exterior of the container frames were clad with everything from metal, glass and wood to finishes of many hues. Potential architectural clients (including those who didn't realize they would become clients following the event, one assumes) were able to browse this library of building materials and finishes while they were in use. Inside, the center was also filled with information for professionals and non-professionals alike.

01 – 03
Floor plans and elevations detailing the layout of materials and finishes on the information kiosk interior and facade.
04, 07
The kiosk was a patchwork of architectural products and a pile instead of a neat stack of container frames of various volumes.
05, 06
The structure hosted a product library and a small terrace cafe overlooking the water.

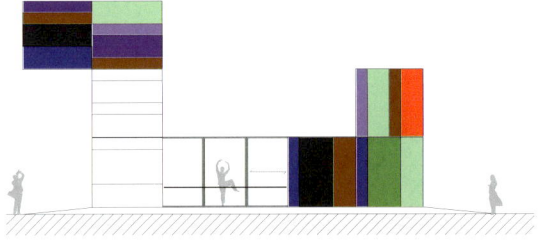

02

01

03

05

- 160 -

06

07

Project
BED BY NIGHT
Architect
HAN SLAWIK

Project location Hanover, Germany
Estimated use   Social service centre
Container type   Building container/freight container

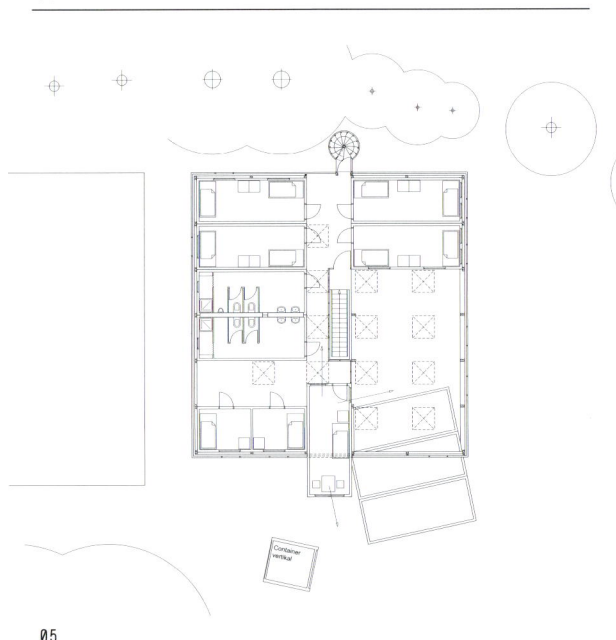

Slawik rehabilitated a single-story structure, assembled from 14 construction site containers, to provide temporary lodgings and social services for street children. As part of the conversion, which will be used for up to 10 years, existing containers were reused and additional modules added. The container ensemble, which Slawik dubbed a "container village", is housed in a protective and insulating premounted industrial glass shell. Slawik placed the new two-story, 361-square-meter space on a frost-proof foundation made of large prefabricated concrete slabs, giving the wooden frame visible joints and attaching the glass facade via explicitly exposed fixtures, to ensure that all structural components were clearly visible. The architecture is meant to give the kids who use it a literal example of the term "support": the conspicuously jointed building serves as a metaphor for the social "loadbearing" that the street children are being taught inside its walls.

01, 02
Slawik painted building and freight containers in primary colors and arranged them in two stories.
03
The "container village" is visible inside an industrial glass shell.
04, 05
Floor plans for ground and first floors.
06, 08, 09
The glass shell is insulating and guards a front courtyard.
07
An exploded view of the containers within the wooden frame and glass shell.

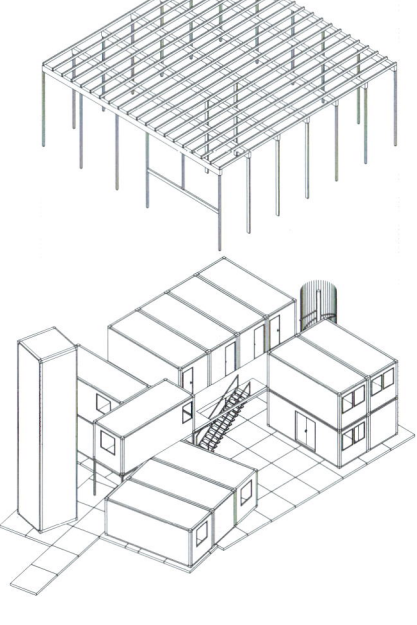

Project
R4HOUSE
Architect
LUIS DE GARRIDO

Project location   Various
Estimated use   Private residence (prototype)
Container type   Freight container

01 – 03
This trade fair prototype for sustainable housing included two domestic spaces built from six repurposed freight containers.

- 164 -

Luis de Garrido's R4House was named for the "Four Rs" that give the architect his inspiration: Recycle (the use of recyclable materials); Recuperate (the recovery of materials); Reuse (of building materials); and Reason (the minimization of costs, waste and energy consumption, the optimization of materials, the consideration of the users' health). An entire pavilion at the Construmat 2007 construction trade fair in Barcelona was dedicated to the topic of sustainable building and prominent among the items on show was Garrido's prototype for a sustainable home. R4 consisted of two housing units: a space with 150 square meters of living area and a small, 30-square-meter apartment. He converted six disused, recycled, 40-foot freight containers into livable indoor rooms and fitted them with facades, with one positioned vertically in order to provide access to each floor. Using containers meant that the house could easily be extended or reduced in size while the energy requirement of construction and disassembly was reduced considerably. All components were designed to be easily demounted and reused and no adhesives or concrete were included; instead elements were screwed, bolted or clamped together. The residential units are so-called "zero-energy houses": the energy required is generated entirely by the house itself from a combination of solar and geothermal power. Planted roofs and automatic control of indoor temperatures also contribute positively to the energy balance. The result represents a refinement of architecture even while the aesthetic appears slightly rough. Garrido describes the container building as having the "beauty of imperfection."

04

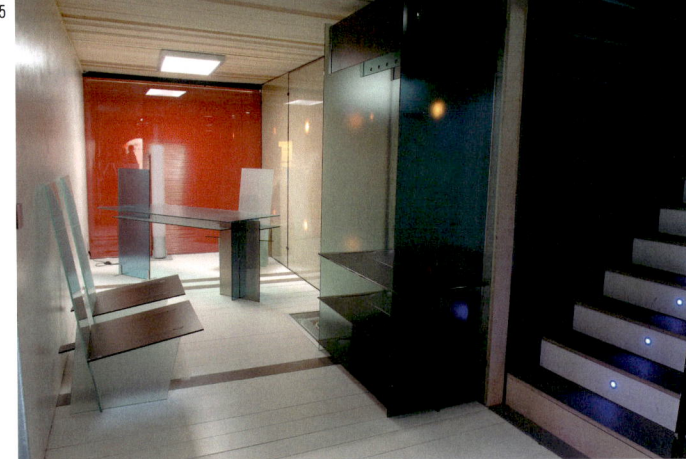

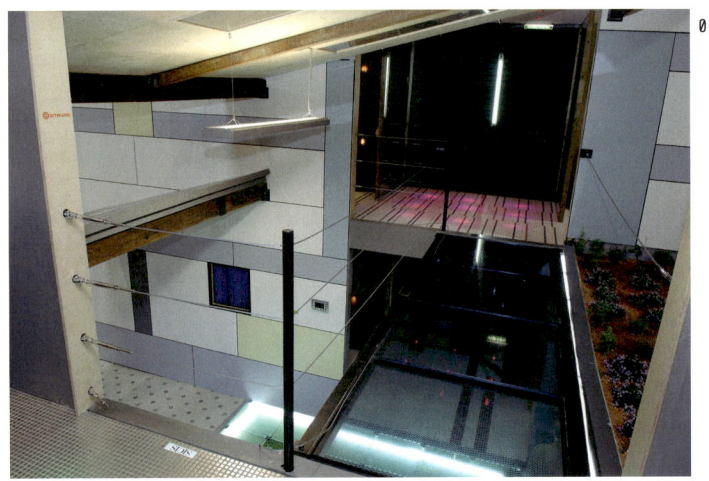

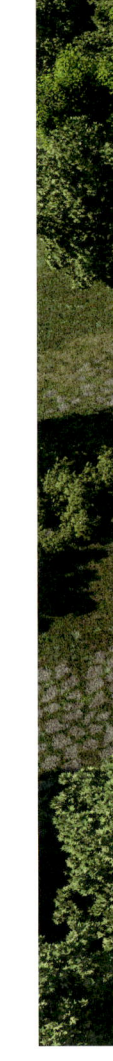

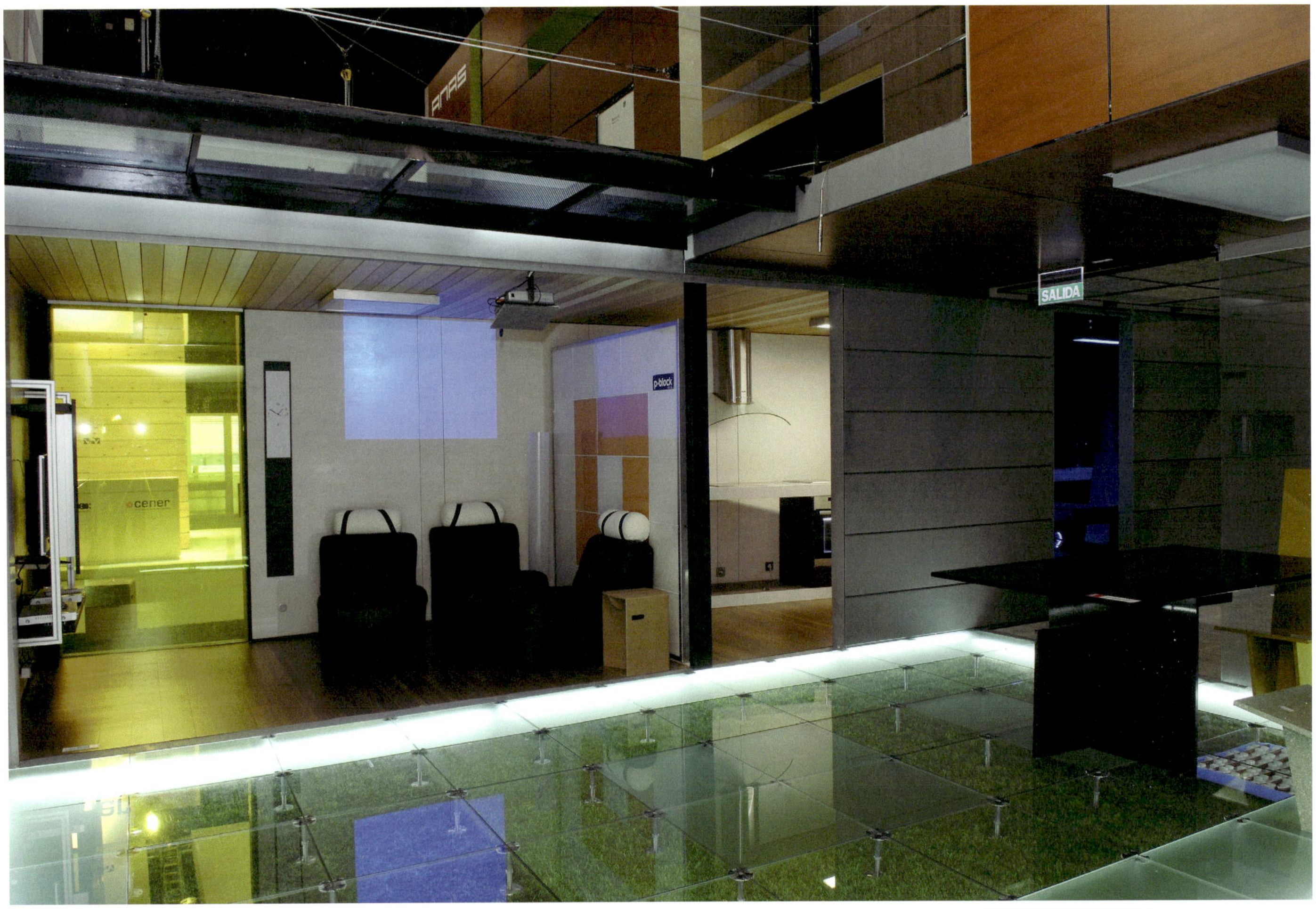

08

05

06

04 – 06
Interiors with glass floors and ceilings.
07
Rendering of a green roof was fitted with extensive solar paneling.
08 – 10
Renderings show the housing in a more natural setting.

07

09

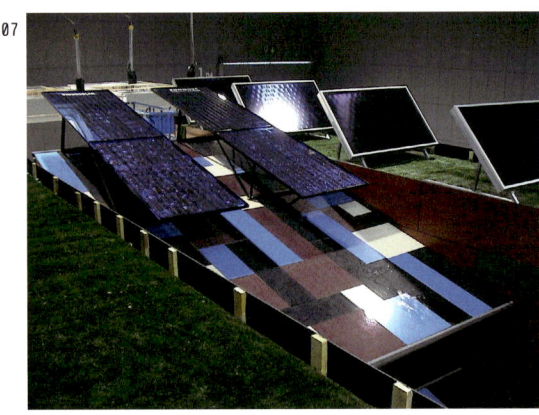

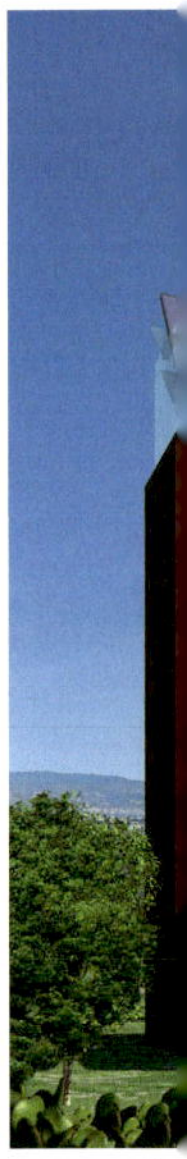

Project
TRINITY BUOY WHARF (CONTAINER CITY I+II)
Architect
NICHOLAS LACEY AND PARTNERS

Project location	London, Great Britain
Estimated use	Housing and studio spaces
Container type	Freight container

01 – 04
The porthole-dotted facades of Lacey's Trinity Buoy Wharf housing with container doors folded down into balconies.
05
East elevation.
06
West elevation.
07
North elevation.
08 [pages 170–171]
The cheerfully painted face of Container City.

On the London Docklands, Nicholas Lacey and Partners used recycled freight containers to build colorful art studios and live-work apartments. Container Cities I and II are situated on Trinity Buoy Wharf, a peninsula at the eastern end of the Docklands. Part of the area's redevelopment into a cultural center for the arts, Buro Happold engineered the projects, which were built in two phases. Container City I was deployed in only three months from fifteen 40-foot containers that were stacked three-high to create 12 workshops, while Container City II added 30 more. The two structures are serviced by a common vertical circulation core and connected by bridges. Two types of windows were used: a large pivoting porthole and a sliding glass door that replaced the standard double doors at the end of each box. Anchored in the open position, these doors now serve as a balcony. The team sprayed the interior of each container with insulation over membrane waterproofing and finished the walls and ceiling in plasterboard. Constructed from 80% recycled materials, the Cities enjoy a long waiting list for rental.

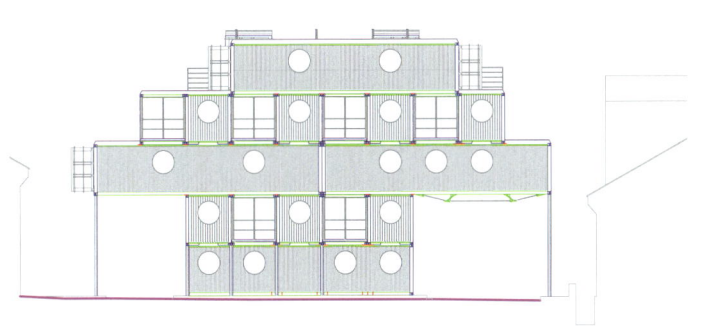

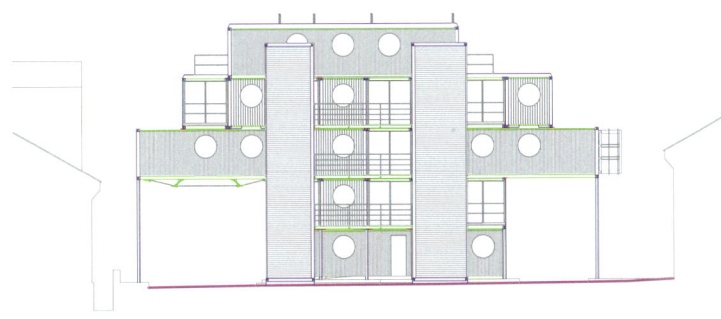

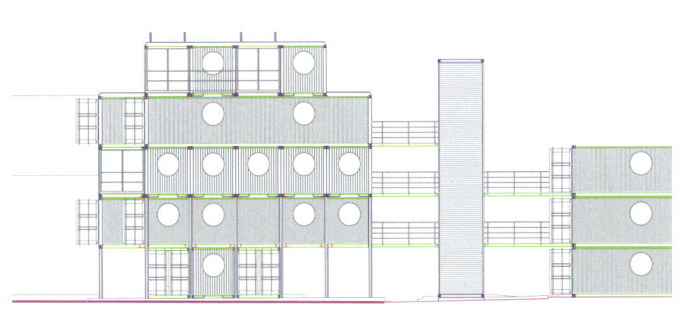

Project
KEETWONEN
Architect
TEMPOHOUSING / JMW ARCHITEKTEN

Project location — Amsterdam, the Netherlands
Estimated use — Student housing
Container type — Freight container

01 – 03
Scarlet framing gives the 1,000 student dormitories a crisp sartorial appearance. In clean rows, the apartments stretch over 17,000 square meters.
04
Mailboxes at one building's entrance and the stairwell that leads to each of the five floors.
05 – 07
Construction photos show that the distinctive red frames are simply an articulation of the red container walls that are the bones of the buildings.

This project is situated in the southeast of Amsterdam, near the highway and a subway station, making it eminently accessible. On a plot of 17,000 square meters, Tempohousing and the De Key housing corporation built a small village of one thousand individual 30-square-meter student dormitories. The entire project consists of six blocks of buildings of varying sizes, arranged to create courtyards that can be used as bicycle parking and social space. The units are stacked up to five levels and sit beside an outdoor basketball field, a supermarket, a launderette, a café, a bicycle repair shop, and several office spaces, built in 34 40-foot containers. To access the living units, staircases and exterior walkways were constructed on the sides facing the courtyard. On the opposite side, each unit has a private balcony. With all its conveniences, the project is also designed to be completely mobile. After five years at this location, it will be transported to a new site in the Netherlands or beyond, since the constituent containers have a Container Safety Convention (CSC) plate and are considered standard freight on container vessels that sail around the globe. All of the living spaces were rented out early on in the project; subsequently the waiting list for rental has grown to over 1.5 years. Keetwonen, which was most likely the largest container village in existence at the time, won a Funda Award in 2006 for "Best executed innovation in construction".

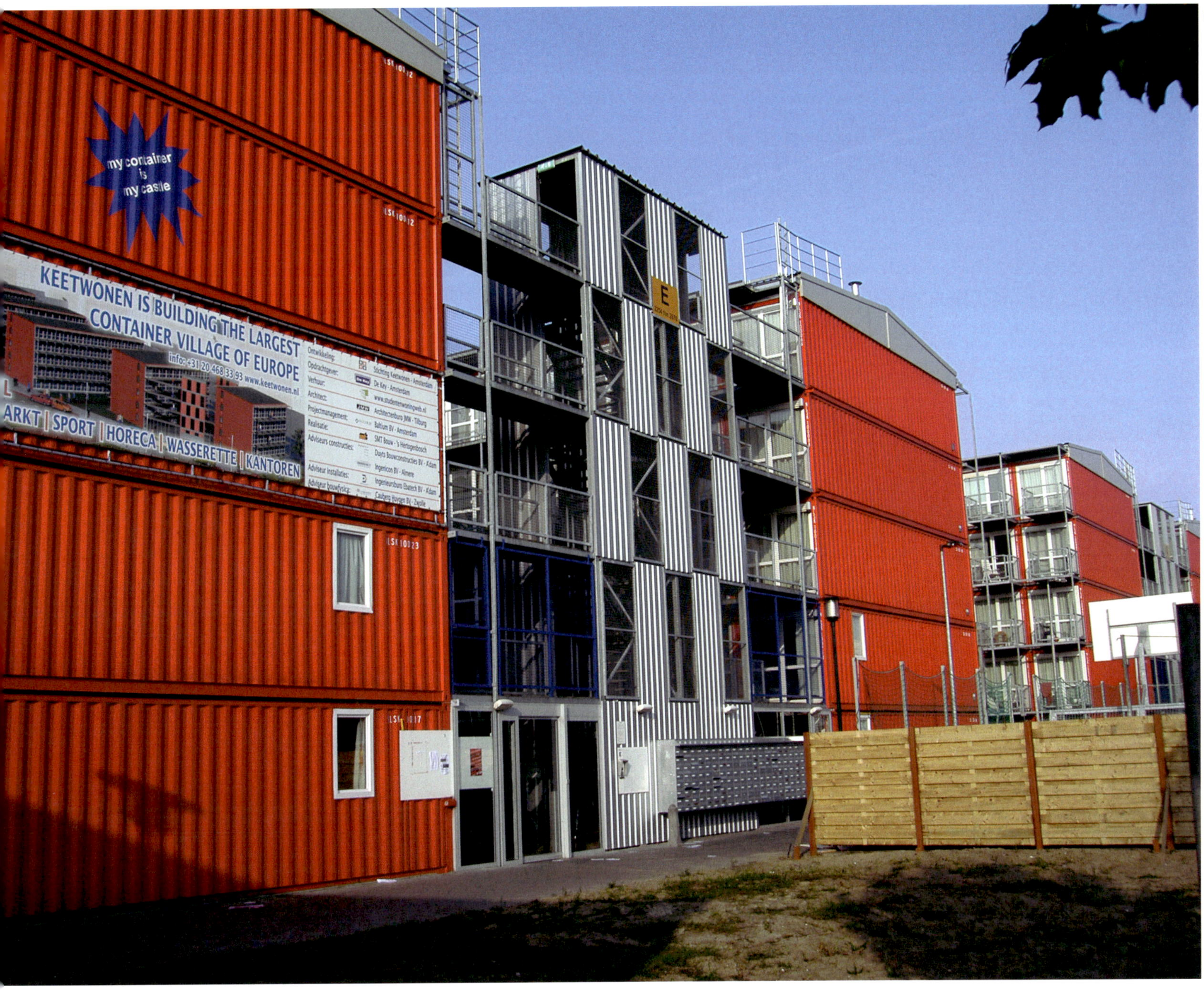
my container is my castle
KEETWONEN IS BUILDING THE LARGEST
CONTAINER VILLAGE OF EUROPE
ARKT | SPORT | HORECA | WASSERETTE | KANTOREN
04

05

MICHIELSENS
06

07

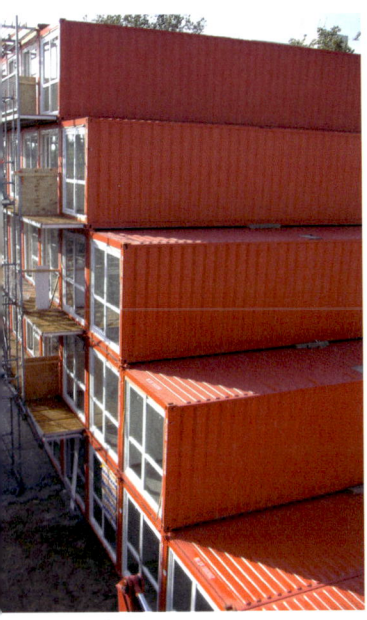

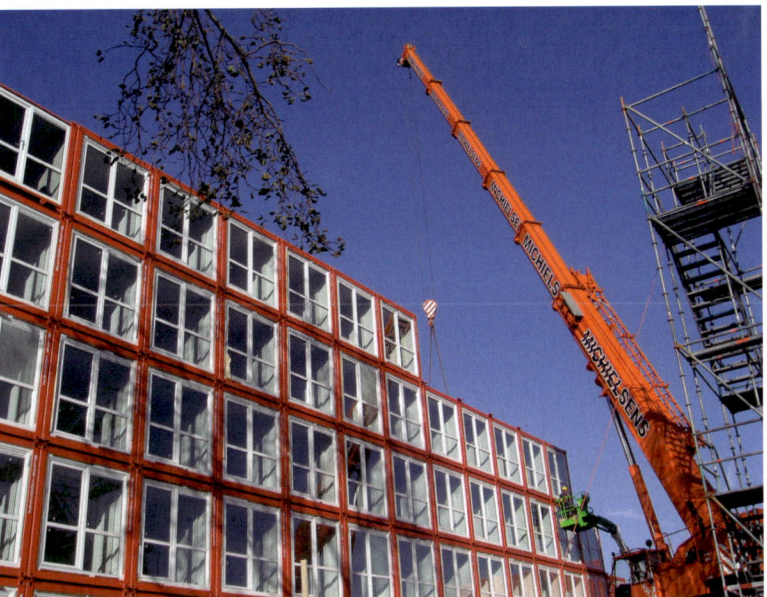

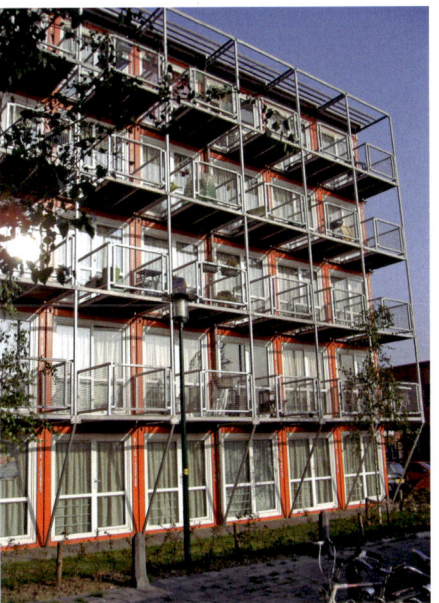

Project
RAINES COURT
Architect
ALLFORD HALL MONAGHAN MORRIS

Project location Stoke Newington, London, Great Britain
Estimated use   Residential housing
Container type   Building container

Raines Court is an £8.9 million building whose design and construction served as an investigation into the potential of off-site volumetric construction. Raines was also the first housing corporation-funded modular housing scheme in the UK. The client, the Peabody Trust, asked AHMM to create high quality flats for sale that would include 41 two-bedrooms, 11 three-bedrooms, one single bedroom, and eight live/work units over three separate blocks, all linked by a central circulation core. The architects had the dwellings manufactured as fully-fitted modules off-site, to factory standards, focusing on maximizing the size of the modules while minimizing the number of components being constructed off-site. The apartments incorporate a fire escape strategy developed in response to the module dimensions; they minimize circulation while maximizing useable area. South-facing covered balconies and private courts that branch off an open walkway are incorporated into the modules, which because they are well-insulated, have low energy requirements and a diminished cost in use. One interesting result of the prefab work undertaken was the record it produced, articulating a number of potential long-term benefits of modular construction including: diminished waste and increased recycling, the greater predictability of the construction process, reduced transportation and disruption of local environments, superior quality control and maintenance, reduced construction time, and increased safety, along with most simply—more cost-effective building. It is worth noting that the Raines Court flats were all sold within weeks of being put on the market.

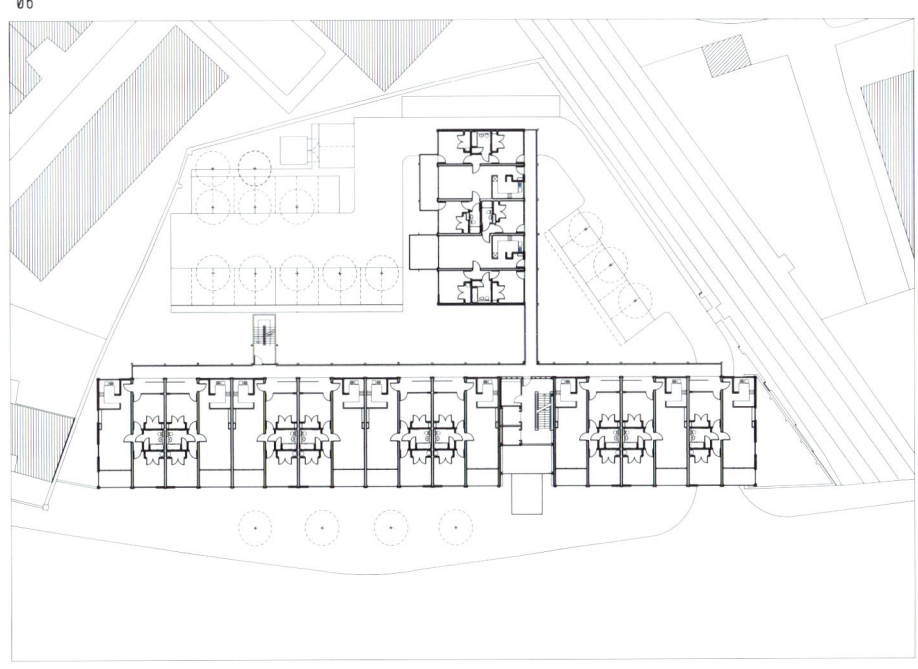

01
Three blocks of prefabricated flats are linked by a central circulation core.
02
Balconies.
03
Because the units were manufactured off-site, the architects were able to maximise their size while reducing the number of components.
04
Apartment interior.
05
South-facing covered balconies.
06
Second floor plan.

Project
QUBIC AMSTERDAM
Architect
HVDN ARCHITECTEN

Project location   Amsterdam, the Netherlands
Estimated use   Student housing
Container type   Building container

This temporary student accommodation was designed to appear permanent. Located on the former Amsterdam's Houthavens dockland, the structure has altered its formerly disheveled surroundings dramatically. The development includes 715 dormitories, 72 apartments, a ship also converted into dorms, artists' ateliers and a variety of restaurants and bars. HVDN stacked containers and sandwiched them between two prefabricated, flat structural planes to achieve a threefold benefit: this ensures that the construction is robust, allows for the creation of slightly elevated verandas—formed by projecting the floor slab beyond the perimeter of the building— and adds to the permanence of its appearance. By removing a number of the containers, the corridors receive sufficient light and ventilation. The facades consist of deep, story-high, molded plastic panels that contain a variety of window openings and colored Plexiglas, amplifying the texture and vivacity of the composition. In order to create a village-like atmosphere, the housing was ranged around three open areas and two courtyards that serve as social space for students.

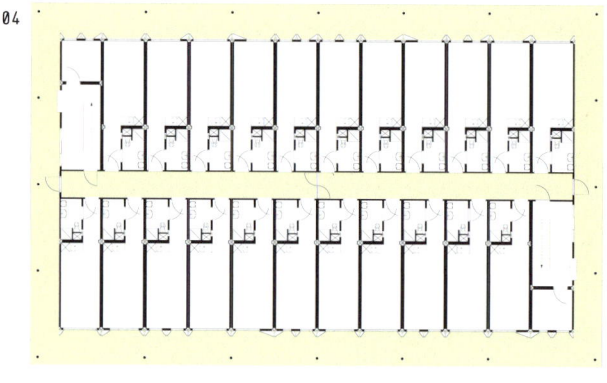

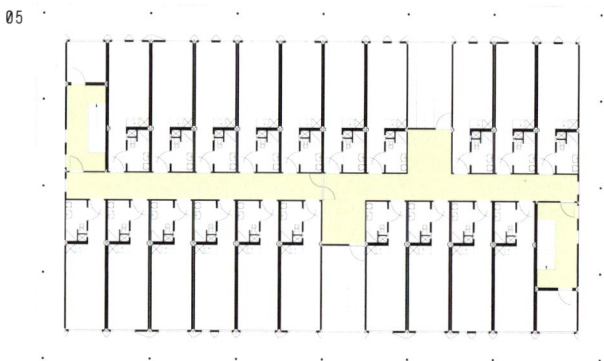

01
The project includes hundreds of dorms (some on a boat), 72 apartments, artists' studios, cafes and bars.
02, 03
The temporary lodgings were designed to feel permanent, even though they were made from building containers and are only needed for about ten years.
04, 05
Ground and first floor plans.
06, 07
Irregular window shapes and orientations combined with vividly colored glass balconies in the white faces of the buildings make the project distinct.

Project

CANCER CENTER AMSTERDAM

Architect

MVRDV

Project location	Amsterdam, the Netherlands
Estimated use	Hospital
Container type	Building container

01

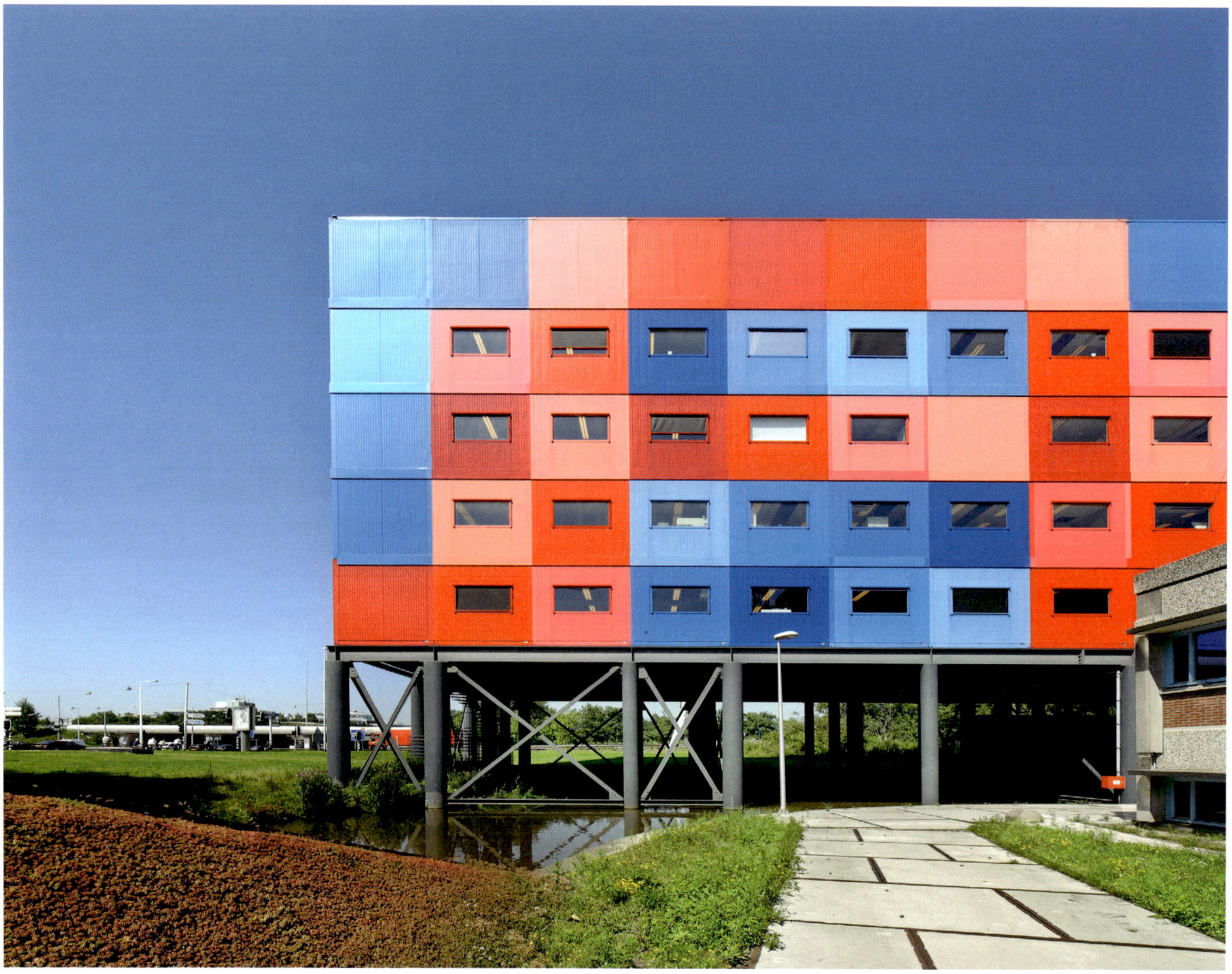

03

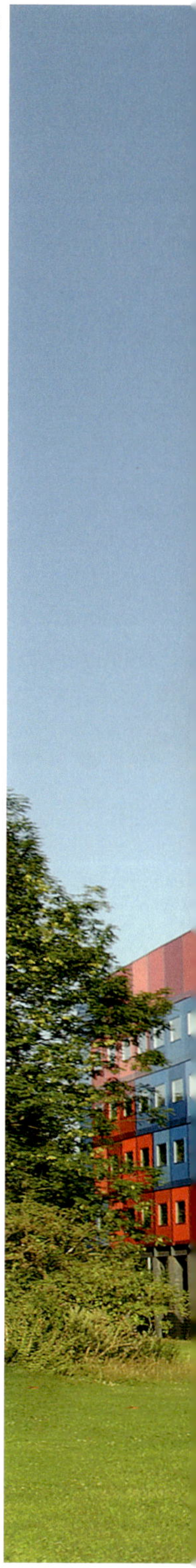

02

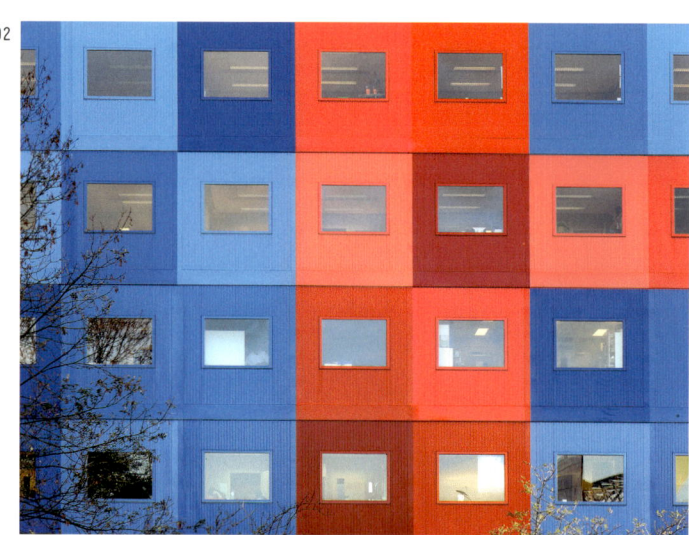

01
MVRDV built vertically on the small site to create a temporary cancer clinic from container-like boxes.
02
The red and blue pixels of the facade emphasises the seemingly modular nature of the building.
03
Beside a major highway, the architecture becomes advertising for the hospital.

During the rebuilding and enlargement of this Amsterdam cancer research and treatment clinic, a part of the Antony van Leeuwenhoek Hospital, Rotterdam-based architects MVRDV designed a temporary institute that could be erected during the construction process. The architects envisioned the temporary structure as a series of containers stacked seven stories high, with a floor space of 1,425 square meters. Located on a small site next to the A19 motorway, it was obvious that the project required a vertical architectural solution. The designers viewed the clinic's location beside one of the country's busiest highways as an opportunity instead of an obstacle. To make the Cancer Center more high profile, they painted each container in various shades of blue and red that assembled together, make the view along the A19 better for everyone's health.

Project
WISMAR TECHNOLOGY AND RESEARCH CENTER
Architect
JEAN NOUVEL
WITH ZIEBELL & PARTNER

Project location	Wismar, Germany
Estimated use	Technology and business center
Container type	Container look

01
The container-like boxes making up the building are neither freight containers nor building containers; rather, they allude to the commercial activities of the nearby port.
02
A view of the peninsula on which the centre is located.
03
The colored volumes on the roof can be seen from afar.
04
An elevation, floor plan and colour scheme.
05 [pages 182–183]
Seen from the water, the centre blends in seamlessly with the surrounding port.

As part of the conversion of the former Wismar port facilities, the TFZ Technology and Research Center offers a creative center for research and business directed primarily at start-up, technology-oriented companies. Because the complex sits at the tip of a peninsula, the colored containers that unevenly crown its roof are visible at a great distance. The two-story, 5,500-square-meter structure features a U-shaped layout that follows the line of the quay. Inside, Nouvel and Ziebell's plan establishes the strong presence of the nearby water: room-height glazing on the ground floor allows an almost seamless transition between inside and outside. On the cantilevered upper floor, visitors float above the harbor. A series of container-like structures placed over the ground floor that house offices, a conference room and spaces for services, are bespoke objects anchored to the flat roof, painted in various colors, and stacked either one or two stories high. However, these boxes are not real freight containers; rather they comprise conventional building components that merely give the appearance of stacked containers. Certain details correspond to the structure of real containers: they are covered with trapezoidal sheeting and have imitation corner fittings, but their width, at 2.85 meters, is tailored to the structural grid of the building instead of standard container dimensions. They cannot be shipped, dismounted or exchanged. Instead, Nouvel's choice of materials establishes visual continuity with the industrial port in order to establish an aesthetic link to the historical site.

03

04

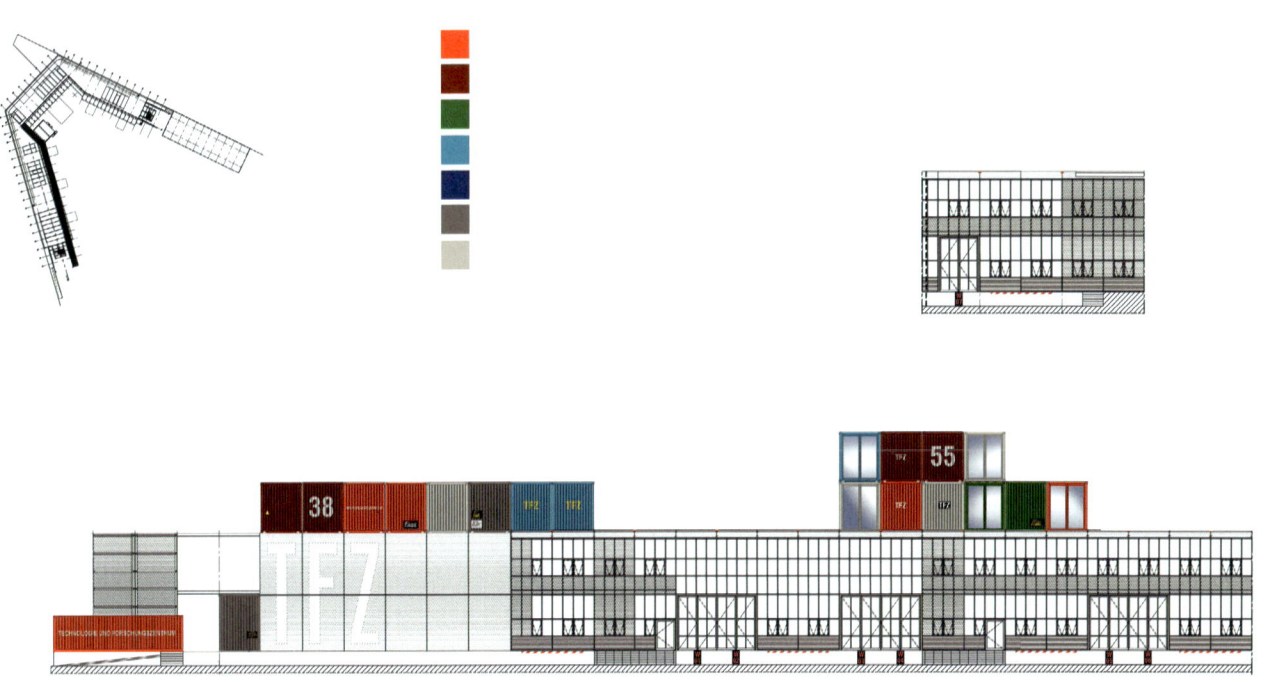

- 181 -

Project
CAMPUS
Architect
HAN SLAWIK

Project location   Almere, the Netherlands
Estimated use   Building laboratory
Container type   Freight container

01

02

03

04

05

- 184 -

01
The interior and view through its floor-to-ceiling (literally) window.
02, 03
Interior details.
04
Axonometric drawing.
05
Floor plan.
06, 07
Four upended containers support a fifth to form a tower.
08
To maintain the loadbearing strength of the containers, Slawik cut only minimally into the boxes to create windows and doors. Between the volumes, he made a light-filled conservatory under large swathes of glass.

Slawik designed a temporary home that would stand for five years in Almere as part of the local government's Temporary Living competition. Campus was one of only 17 entries selected to be built and was the only project constructed from steel shipping containers. Campus is considered by some to be the spark that initiated container architecture in Europe. The architect placed four 20-foot steel shipping containers vertically on the flat polder landscape to support a fifth that served as a small tower and determined the separation of the four boxes below it. The space between them was enclosed by a single glazing and could be used as a conservatory. Because containers are designed to support excess loading, weakening them by cutting openings into them during their conversion into housing was feasible. Nonetheless, Slawik cut only small square openings into the trapezoidal sheeting walls in order to avoid having to use additional reinforcement. Door openings were cut from the walls, fitted with frames, and then the original cut-outs were used as doors. The modular dimension of 2.26 meters, borrowed from Le Corbusier, formed the basis of Campus' interior planning grid, however, the 20-foot containers, by their nature, already lent an outward impression of order to the assemblage.

Project
MYSTERY CUBE
Architect
HAN SLAWIK

Project location   Hanover, Germany
Estimated use   Exhibition space
Container type   Building container

This temporary pavilion can be assembled, disassembled and reassembled to serve as exhibition space. Since no floor space could be allocated in the forum of Hanover's State Museum for educational projects related to The Mysterious Bog People exhibition, Slawik designed an external space housed in recycled building containers that were not immediately legible as such to the public. The containers were first stacked, surrounded by standard steel-tube scaffolding and then dressed in a black fiberglass textile. The result was a 7-meter cube rotated in such a way that it stood out clearly amongst the surrounding city fabric. Unlike most traveling show pavilions, Mystery Cube's standardized components facilitate its international transport by land or sea.

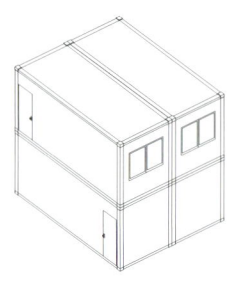 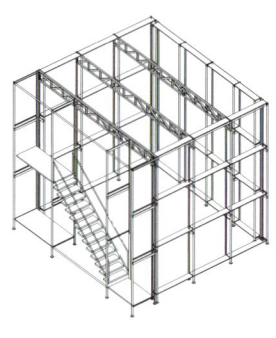 

01 – 03
Mystery Cube deployed in a Hanover plaza, supplementing exibition space at the State Museum.
04
The structure consisted of three layers.

Project
DERCUBE
Architect
LHVH ARCHITEKTEN

Project location   Freudenberg, Germany
Estimated use   Exhibition space
Container type   Building container

01
The temporary exhibition space became sheer at night through a textile cladding.
02, 03
During the day it looked opaque and monolithic, since fabric walls, which were later recycled, hid the 36 containers making up the three-story structure.
04
Entrance to the building.
05, 06
Floor plans.

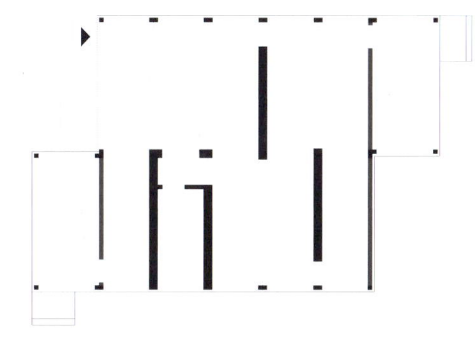

05
06

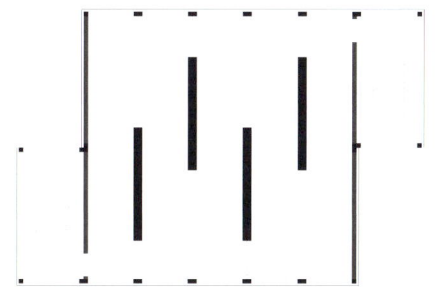

derCube was a three-month extension to an exhibition that outgrew its event space in Freudenberg in the south of North Rhine-Westphalia. A three-story cube with approximately 400 square meters of floor area was assembled on the market square in the town's historical center using 36 containers. The modular cube was distinguished by its unusual textile facade, which was made up of polyester strips coated with PVC. As it wrapped the underlying container grid, it disguised the modularity of its construction, creating the impression of a seamless and monolithic cube. When the exhibition was over, the containers were used again as office space, and the cloth facade was recycled. Through this project, the architects explored the ways in which a town can be shaped and altered by temporary structures. They discovered that the potential of certain locations can be revealed through temporary interventions; at the same time, new interpretations of existing structures also become imaginable.

Project
SEVEN-DAY ARCHITECTURE
Architect
LHVH ARCHITEKTEN

Project location   Munich, Germany
Estimated use   Showroom
Container type   Building container

01
A textile made this grid of containers look uniform.
02
Some of the building containers used were stripped to their frames.
03
During the day, the pavilion looked opaque; as night fell, it revealed its layers.
04
Views in section.

This trade fair stand by LHVH featured 16 containers in the shape of a grid with additional boxes stacked to create an interior building of 27 containers ranged over three floors. The interiors included reception and café, stairs and a roof terrace. Another assemblage of 21 containers surrounded the inner booth with some stripped to their steel supporting skeletons, leaving a three-dimensional structure of x, y, z coordinates. After assembly, the components were sheathed in a vinyl mesh printed in a color that created a peaceful atmosphere. During the day, the stand's exterior was opaque—a quiet, impenetrable, minimalist box—as night fell, the shell became sheer to reveal its sculptured interior. Following the seven-day event, each container was returned to its original function.

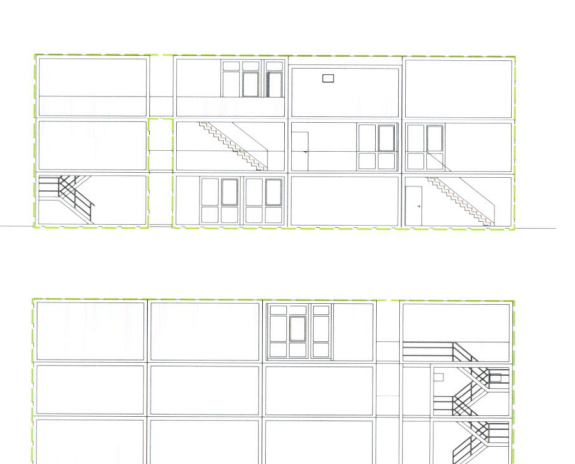

Project
50,000€ MODULAR-HOUSE
Architect
MM+P / MEYER-MIETHKE & PARTNER

Project location — Leipzig, Germany
Estimated use — Office and private residence
Container type — Container frames

01

05

02

03

04

01 – 03
This modular steel home was made from modified building containers with copious incisions that allowed designers to insert windows all around.
04
Looking down from the first floor.
05
The double-height interior and views that connect the home to its surroundings.

A 70-square-meter, cube-based modular home made of steel and glass for less than €100,000? This kind of affordability was a product of mm+p architects clever choice to combine a high degree of prefabrication and standardization with a super-efficient layout. The designers worked with a limited range of material formats—incorporating only two glazing sizes, for example—and a simple flat roof made from steel sheeting that doubles as a water collection basin. The design places living, cooking and eating areas on the first floor in such a way that they seem to form a single unit with the garden and terrace, even when the windows are closed. It is the glassy transparency that dissolves the boundaries of the house almost completely.

Project
TEMPORARY ACCOMMODATION FOR 1001 DAYS
Architect
MEYER EN VAN SCHOOTEN ARCHITECTEN

Project location   Amsterdam, the Netherlands
Estimated use   Office space
Container type   Freight container

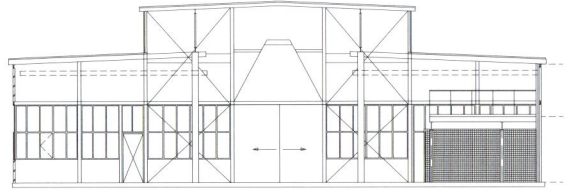
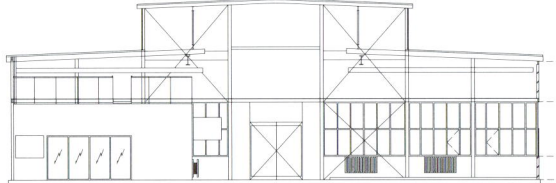

A large factory hall on the former site of the Stork engineering company was converted over a period of two months into an office by the architects who would inhabit it for the next few years. Blue freight containers were stacked in the vast 800-square-meter interior. The re-used freight containers not only formed rooms within the room but helped to define the space as a whole. The containers also had the virtue of gracefully expressing the temporary nature of the accommodation. Meyer en Van Schooten Architecten thrust a container through the outer wall to create the entrance. On the longitudinal axis of the shed, where it was tallest, they placed two structures, one consisting of four containers and the other of two. The largest block, in which the ends of alternating containers were replaced by glass, contained the secretarial office and a conference room. The smaller block contained the network server and archives. Elsewhere, container clusters housed a model workshop, the lunch area, kitchen, sanitary facilities, a library, and management offices. The units were fitted with sliding walls and ceilings made of translucent plastic panels to create privacy and diminish distractions from without, while retaining heat during the winter.

01
The former factory building in which the architects inserted their temporary offices.
02, 03
Freight containers created rooms on the factory floor for everything from meeting space and a model workshop to offices and a library.
04
Some container walls were replaced with glass, sliding walls or ceilings made from translucent plastic.
05
An area for model display.
06
Ground floor plan.
07
Plan for the first floor.

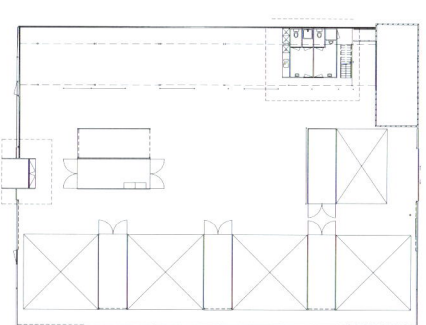

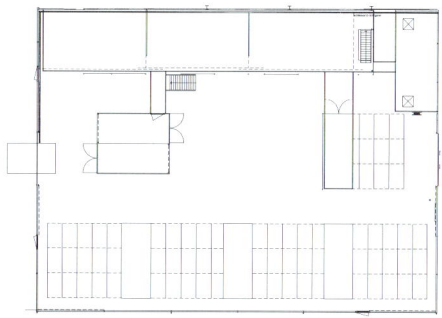

Project
PLATOON KUNSTHALLE SEOUL

Architect
PLATOON AND GRAFT ARCHITECTS

WITH U-IL ARCHITECTS

Project location	Seoul, Korea
Estimated use	Event space
Container type	Freight container

In the upscale Cheongdam area, beside upscale galleries and boutiques, PLATOON KUNSTHALLE Seoul's 950-square-meter space for events, exhibitions, art studios, a workshop, a public library, offices, and a bar and restaurant, consists of 28 reinforced cargo containers that have been stacked and welded together. Graft Architects' framed, plasterboard, mineral wool and aluminium plate interiors are dedicated to hosting art and cultural events, shows and workshops that usually fall beneath the radar, including street art, graphic design, fashion, video art, programming, music, club culture, and political activism. The salvaged containers express this alternative institutional character in a dynamic space made from what have become icons of flexible, responsible architecture in today's "Think Global, Act Local" atmosphere. They can also be rebuilt anywhere, at any time. This is no small virtue.

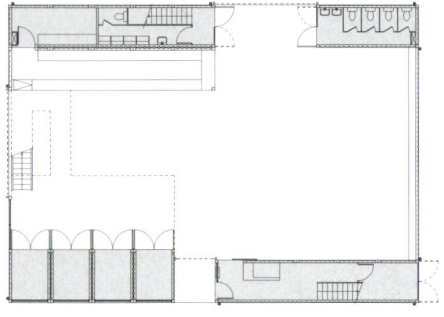
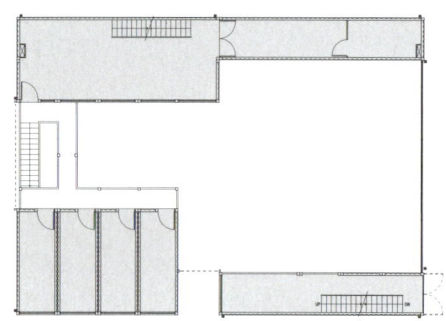
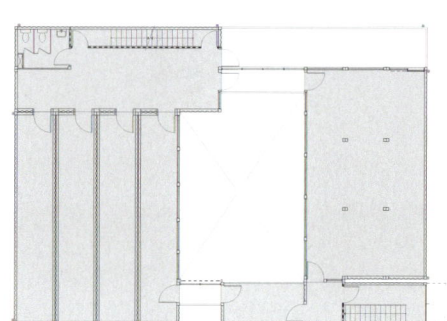

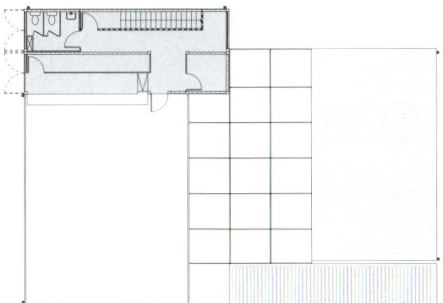

01
Floor plans.
02, 03
Views of the rear of the building.
04
Front facade of the event space.

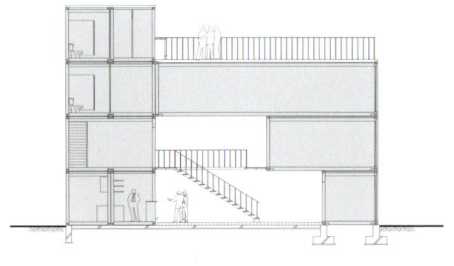
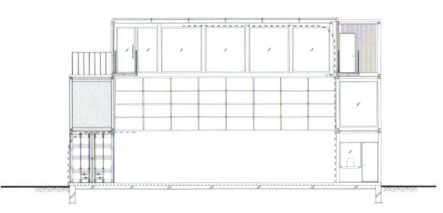
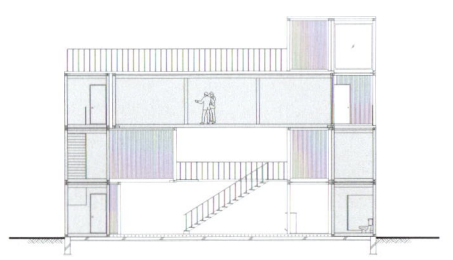
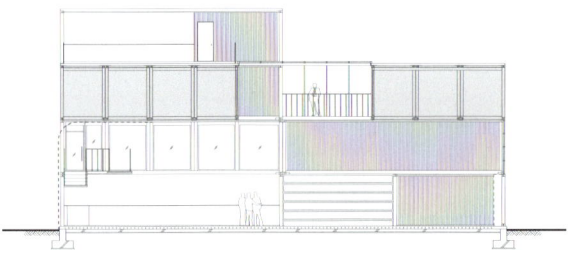

05

05
The building in sections.
06
The glazed ends of containers are juxtaposed with the windowless longitudinal sides of adjacent containers.

07, 10
The main hall.
08
The library.
09
An industrial gate serves
as a side door and window.

Project
VOLVO C30 EXPERIENCE PAVILION
Architect
KNOCK

Project location   Gotheburg, Sweden
Estimated use   Showroom and event space
Container type   Freight container

Stockholm-based architects Knock designed this pavilion to present Volvo's 2006 C30 model to more than 6,000 dealers at the company's annual eight-week convention. Characterizing the car through oppositions—as more emotional than rational, raw yet smooth and having attitude while still being a Volvo—the architects emphasized contrasts, placing unpolished materials, beside clean, white surfaces, for instance. Each area of the event was assigned a distinct audio and visual code, built on these contrasts and built in the shape of the EKG line of a beating heart. Situated in an old boat shed in the rural harbor outside Gothenburg, the pavilion was accessed through a dim, 50-meter tunnel made from containers illuminated with LED strips only. In the main hall, dealers found the broader walls lined with rusty containers stacked three high whose shorter ends served as projection screens. A 150-meter white conveyor belt wound through the room bearing hors d'oeuvres, along with objects identified as being of particular interest to the model's youthful target demographic. The participants were never allowed to see the car. Instead, every element of the pavilion interiors reflected the C30 in some way, giving the expectant audience all the clues needed to piece together an image of it in their minds.

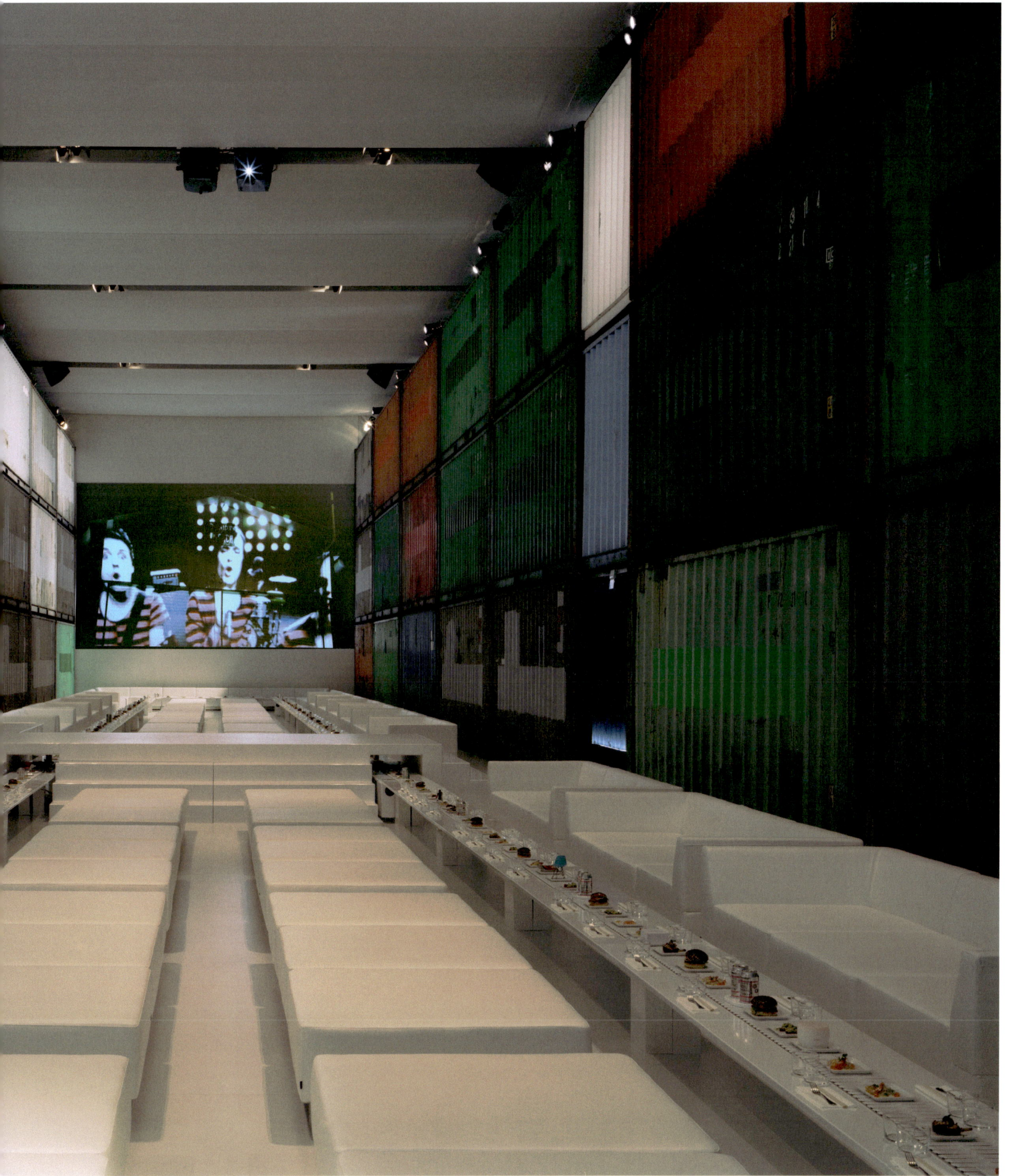

01
Visitors entered through a 50-meter tunnel of containers.
02
Raw corrugated metal was paired with immaculate white. A conveyor belt carried hors d'oeuvres.

Project
NOMADIC MUSEUM
Architect
SHIGERU BAN ARCHITECTS

Project location: New York City, New York, USA
Estimated use: Exhibition space
Container type: Freight container

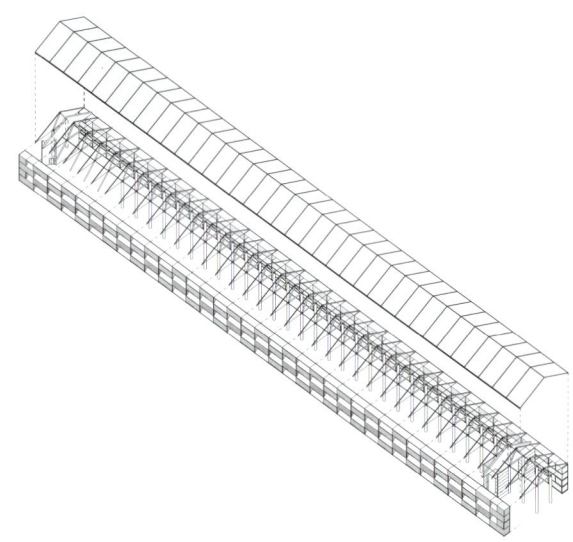

01

01
Exploded view of 148 containers that formed this floating gallery.
02
The Nomadic Museum greeted visitors on a New York City pier, combining recycled and ecofriendly materials for which Ban is famous.

Tokyo-based architect Shigeru Ban didn't reify the freight container in his temporary nomadic museum; instead, he used it pragmatically to construct the cathedral-like exhibition space. It took 148 multicolored shipping containers to raise the structure at the edge of the Hudson River at Pier 54 in Manhattan, and to demonstrate the architect's command of both materials and sustainable practices. The 3,020-square-meter, four-story floating and movable mammoth featured a PVC membrane roof and walls, as well as plywood walls, columns built from water-sealed paper tube (a feature of many of the architect's most famous works), wood plank decking, and 300 tons of gravel that gave the project an Eastern sensibility.

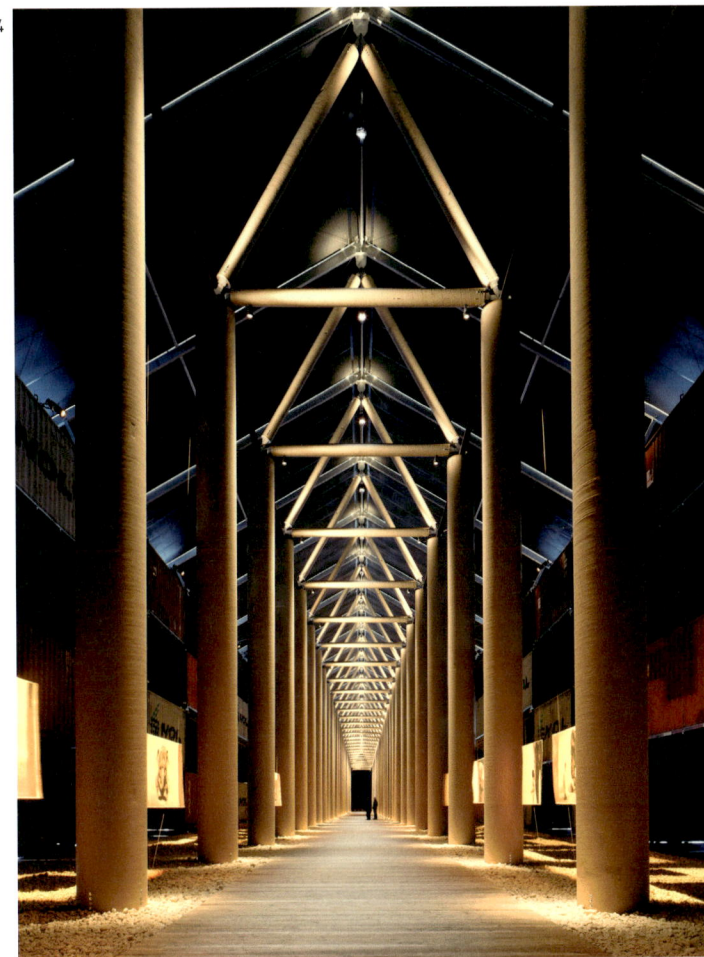

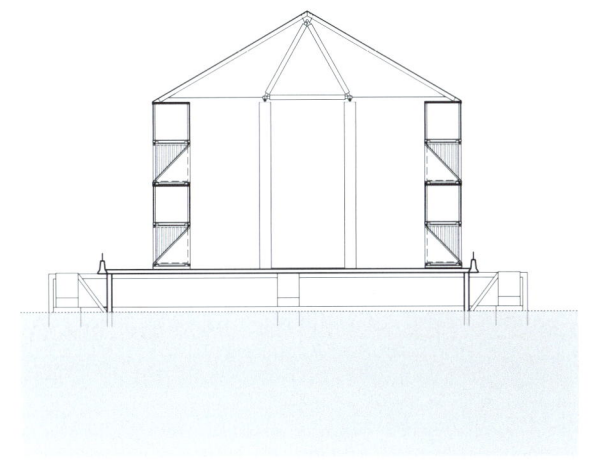

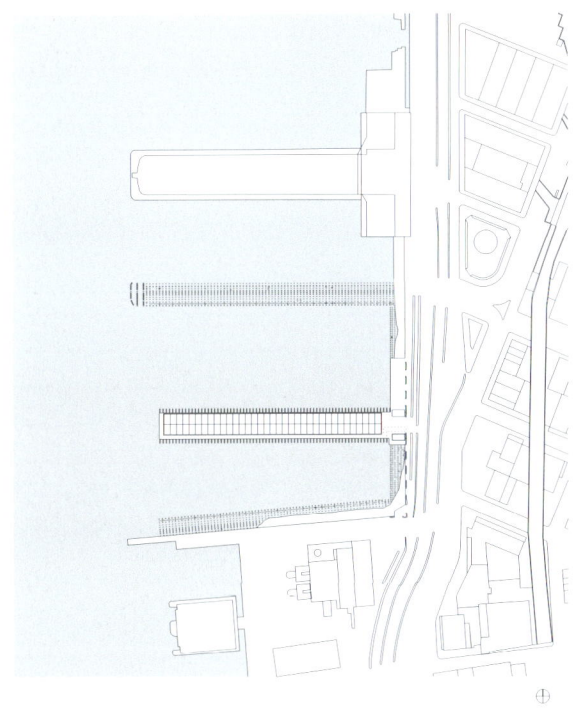

Famous for the use of paper in his architecture, Ban also supended a sheer handmade curtain made of one million pressed paper Sri Lankan tea bags 40 feet above a wooden walkway that begins at the entrance, and in combination with the gravel, tightly controlled circulation through the show. Walls and partitions throughout the space were filled with Ashes and Snow, a multimedia exhibit by artist Gregory Colbert that included 199 large-scale photographs and a one-hour 35-millimeter film. The surreal projected portraits of animals and humans felt at home in the post-industrial cathedral in which narrowly targeted lighting left the edges of the space dim and allowed the corrugated metal surfaces of the containers to recede into darkness.

03
The museum seen against the Manhattan skyline from the Hudson River.
04
The architect's signature paper tubes served as cathedral-scale columns, creating an arcade and naves that limited circulation.
05
Photographs were suspended, floating, between columns above 300 tons of gravel.
06
A view in section: Carefully focused lighting made the raw metal walls fade into darkness, leaving more refined elements in the foreground.
07
Site plan.

Project
PAPERTAINER MUSEUM
Architect
SHIGERU BAN ARCHITECTS + KACI INTERNATIONAL INC.

Project location Seoul, Korea
Estimated use   Exhibition space
Container type   Freight container

Ban's temporary exhibition pavilion built with cargo containers and paper tubes, was created in the Seoul Olympic Park on the occasion of the 30th anniversary of the Korean publisher Design House. Born in Tokyo in 1957, Ban studied architecture at the Southern California Institute of Architecture and the Cooper Union School of Architecture in New York and then worked for Arata Isozaki in Tokyo before founding his own practice in 1985. He began to explore the structural use of paper at different scales in 1985 and in greater earnest began to use paper tubes in the 1989 Paper Arbor in Nagoya. The site of the oddly named Papertainer Museum is encircled by a forest and a roundabout, anchored with a big sculpture at its center. All of the features of the site give it the shape of the letter "D" in plan, which, conveniently, also represents the first initial of the name of the client, which requested that Ban create two types of gallery in the 3,820-square-meter pavilion: one, a multipurpose space and gallery in which to hang large photography and, at the front part of the building, the Container Gallery. The Container Gallery is defined by two parallel walls, each approximately 10 meters high and made in a checker-board composition. Atop the steel-framed walls, like cross beams or a lid over the gallery below, Ban stacked ten 40-foot shipping containers. One container wall is used for exhibition booths and the second as office and storage spaces. The Semicircular Pavilion, also called the Paper Gallery, is composed of two circular walls made of Ban's signature paper tubes which serve as 75-centimeter posts, and support a roof truss that is also constructed from paper tubes.

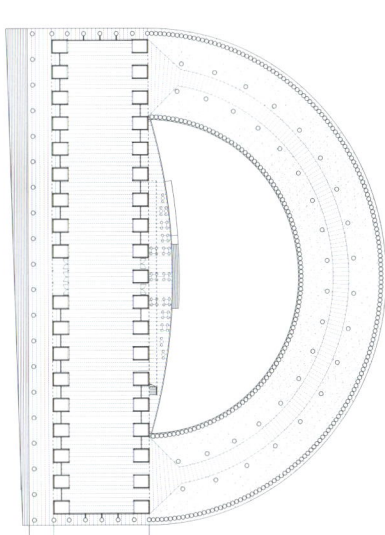

01, 04
The museum facade with colonnades consisting of paper columns fronting a container gallery, office and storage space.
02, 05
The hemispherical plan of the building seen from overhead and in floor plan.
03
Inside, cardboard cylinders, gravel and a wooden walkway suggest the meditative atmosphere of a house of worship.
06 [pages 206–207]
Inside the textured and softly illuminated Container Gallery.

Project
CRUISE CENTER
Architect
RENNER HAINKE WIRTH ARCHITEKTEN

Project location Hamburg, Germany
Estimated use   Cruise terminal
Container type   Freight container, container frames

01
The terminal's trapezoidal, illuminated roof makes a visual allusion to a white canvas sail.
02
The interior recalls a ship deck and the industrial vitality of the port.
03
A translucent foil applied to the container-lined facade makes the surfaces shine at night.

Hamburg-based architects Renner Hainke Wirth Architekten (RHWA) designed this temporary cruise terminal, taking their cues from two familiar elements of seafaring: the traditional overseas cargo container and the large, white canvas sail (which doubles nicely as a symbol for the luxurious, active life that includes shipboard holidaying). The corporate identity of the terminal came from a combination of the building's rough shell, assembled from existing container components, and a vast cantilevered roof that faces the city. A bright tarmac surface covered with colored carpet-like sections recalls a ship's deck and corresponds with the brightness of the shining white roof. At night, the roof resembles an illuminated and floating sheet of canvas while nearby, a grid of vertical container follies also shine through the darkness due to the application of a translucent foil that inspired RHWA to dub them "lighthouses". The containers are painted in a range of maritime shades of blue. In contrast, their interior of the building is lined with a red floor. Like a window on the city, a wide opening in the terminal's facade directs the eyes of arriving cruise ship passengers to the city's skyline and towards Hamburg's renowned St. Michaelis Church.

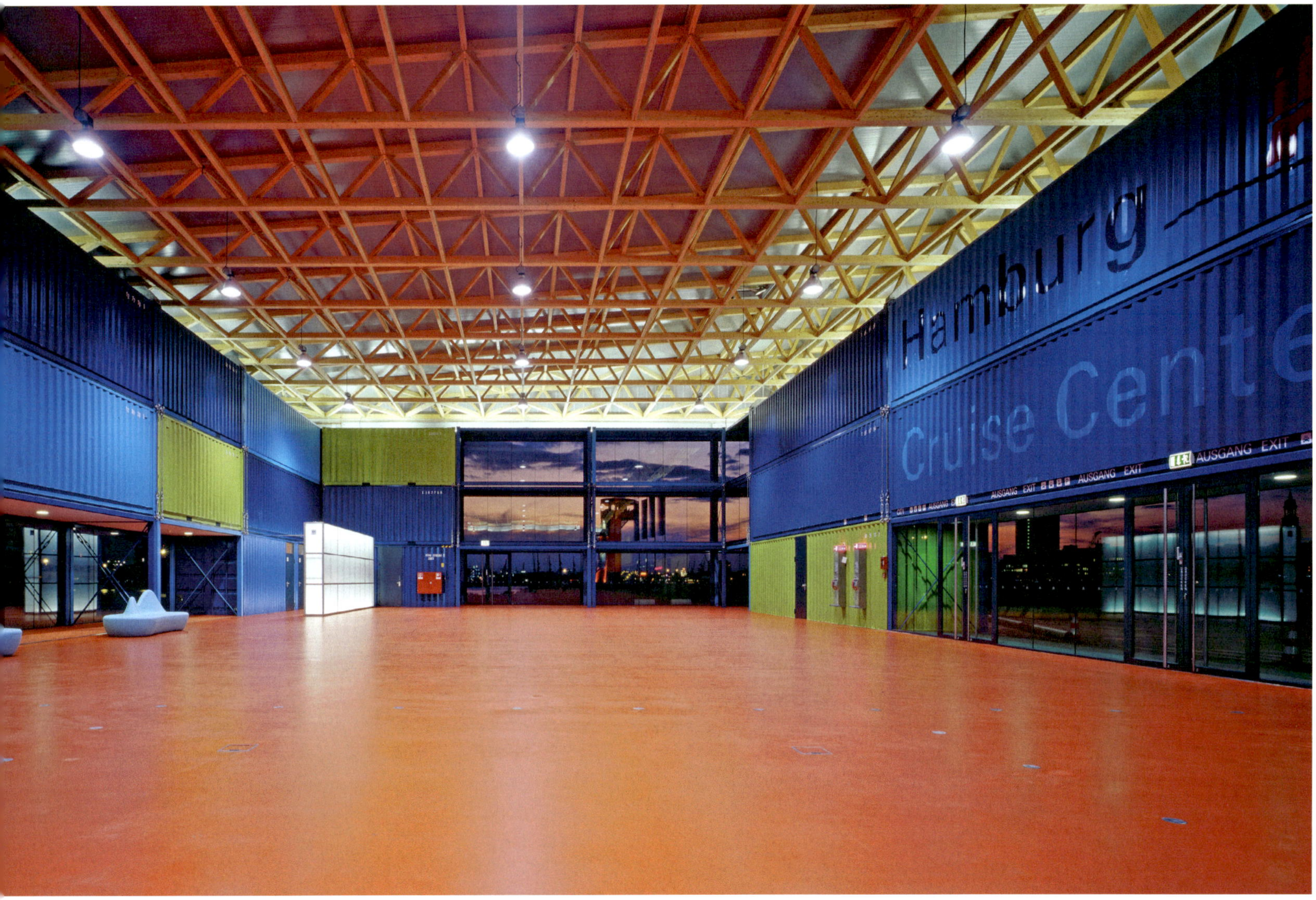
Hamburg
Cruise Center
AUSGANG EXIT

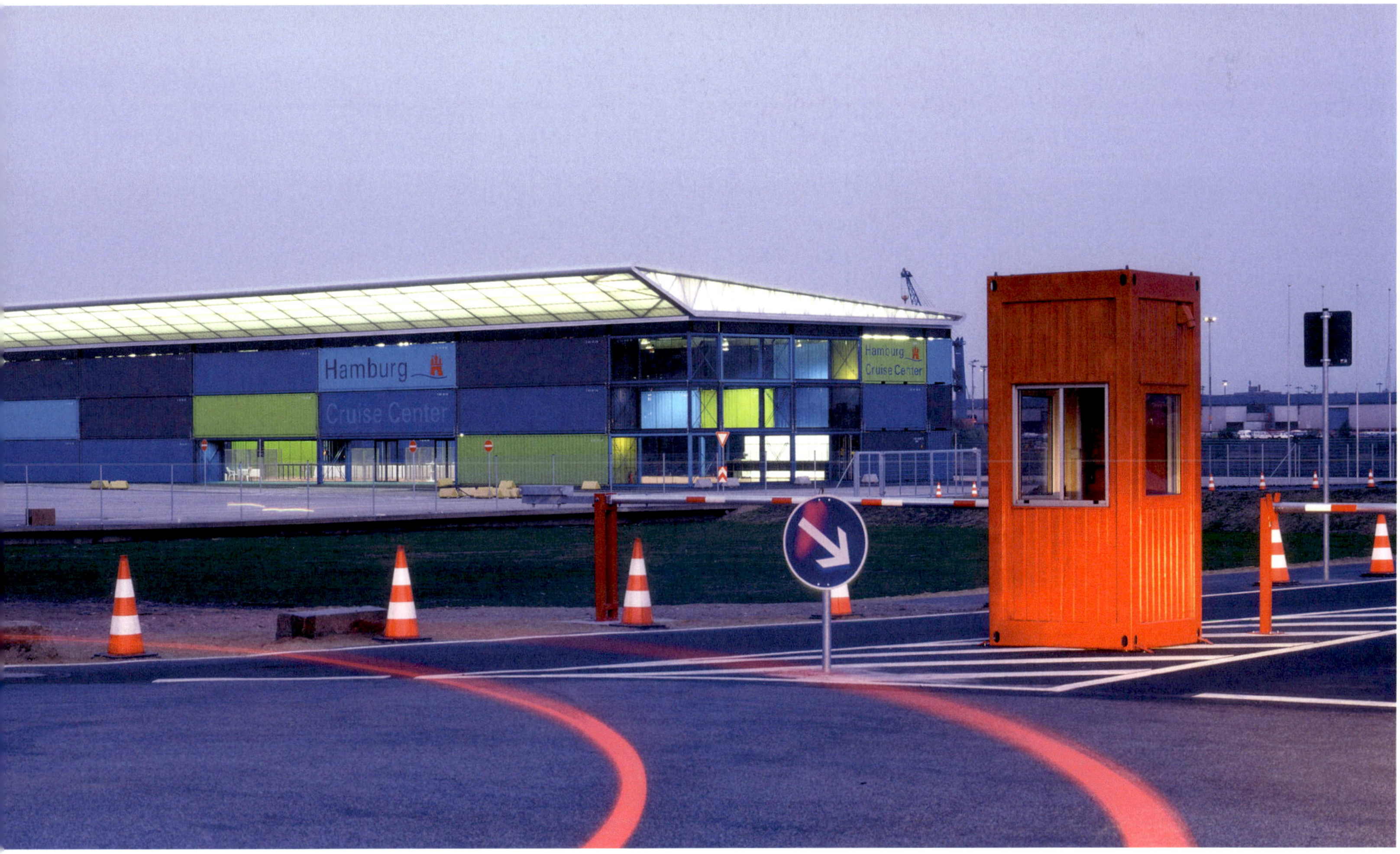
Hamburg
Cruise Center

05

06

07

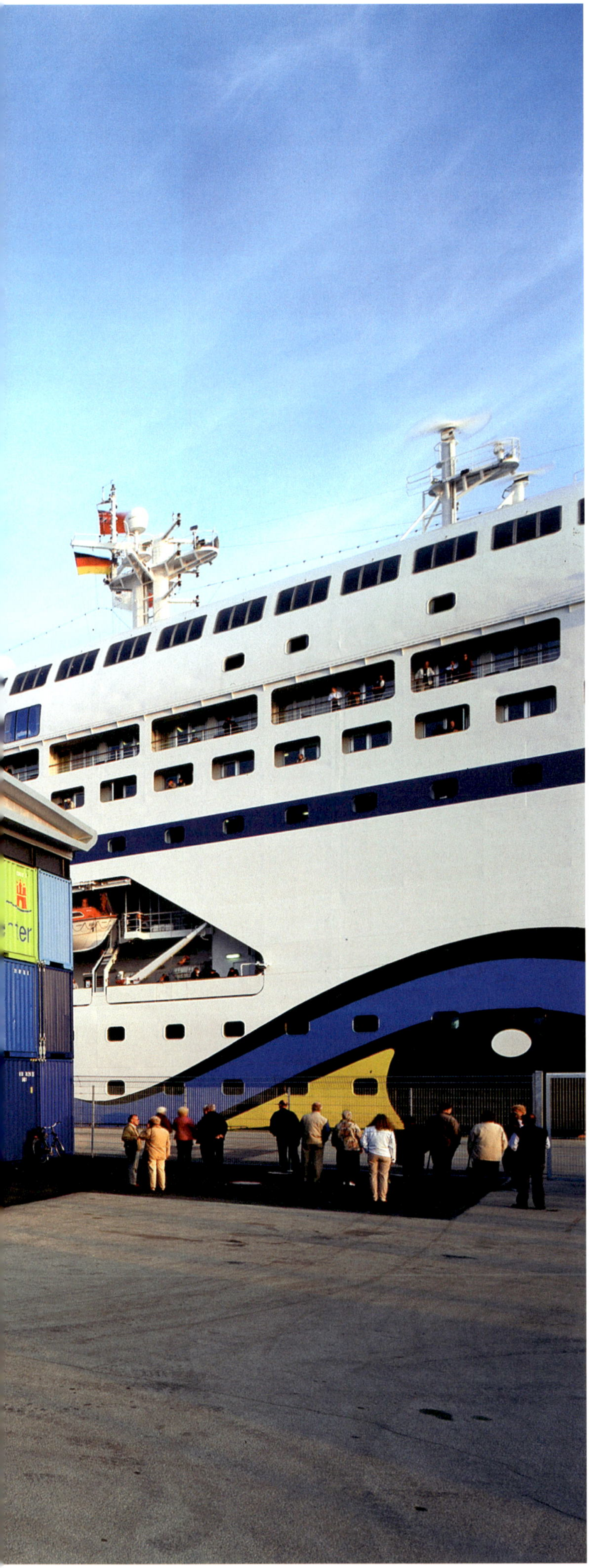

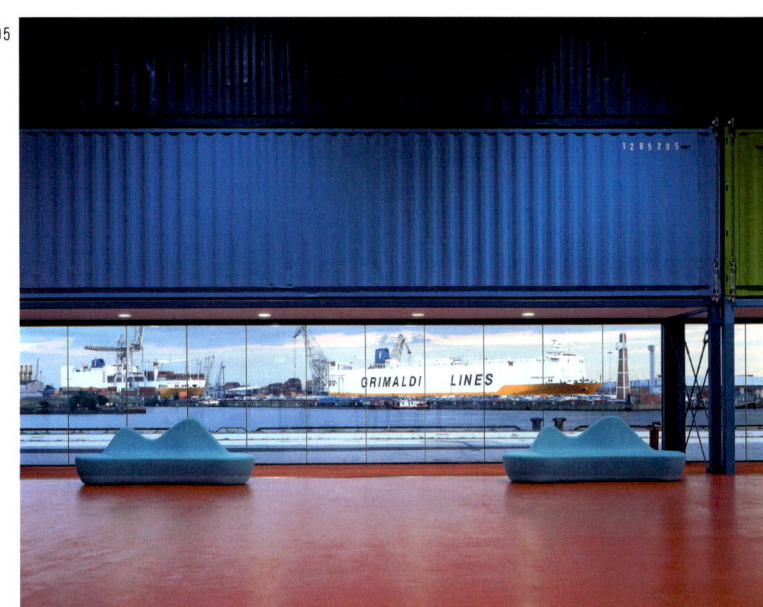

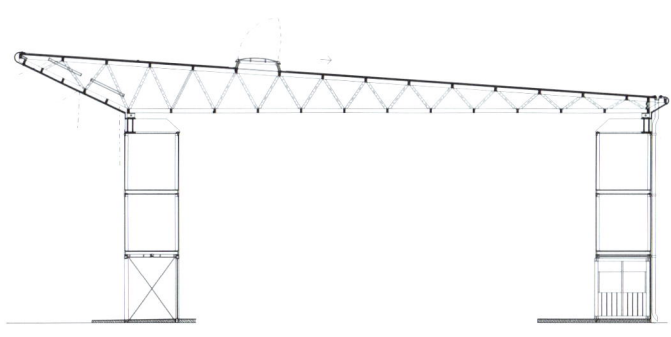

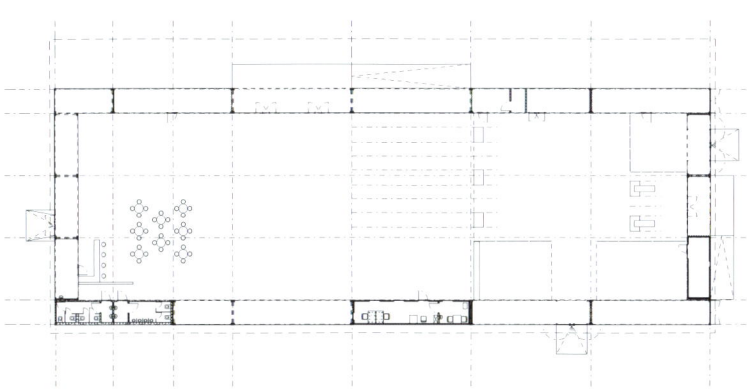

04
The heavy cantilever of the roof is as striking an image as the gargantuan cruise ships that resemble cities floating impossibly on the sea.
05
The ship-shape bareness of the interior and a view to the water.
06
Section with roof structure.
07
Floor plan.

Project
FLYPORT
Architect
WOLFGANG LATZEL ARCHITEKTEN
WOLFGANG LATZEL, NORBERT ZSCHORNACK

Project location   Various
Estimated use   Passenger terminal
Container type   Container frames

This standardized passenger terminal consisting of approximately 400 modules is laid out for 0.5 to 1.5 million passengers a year. The modular concept based on the standardized 20-foot container enables transport via truck, train, ship, and aircraft. Thanks to its modular design, the flyport not only ensures an extremely short realization time; its great adaptability in regard to the respective requirements also allows for flexible usage and individual design. Especially in times of dynamic increase in passenger capacities, the passenger terminal system offers a quick, flexible and individual solution. The flyport is delivered to sites in Europe, Asia, Africa, Latin America, the Caribbean, the Pacific region, and North America. Three standards have been developed: high, medium and basic standard, depending on the demands of the user or operator. High standard is for permanent regional airport operations, medium standard was developed for both permanent and temporary expansion terminals, while the basic design is aimed at the low-cost terminal market.

03

04

01
Exterior view of the passenger terminal with three levels.
02
Interior view of the main hall.
03
View of two-levelled passenger terminal.
04
Building sections.

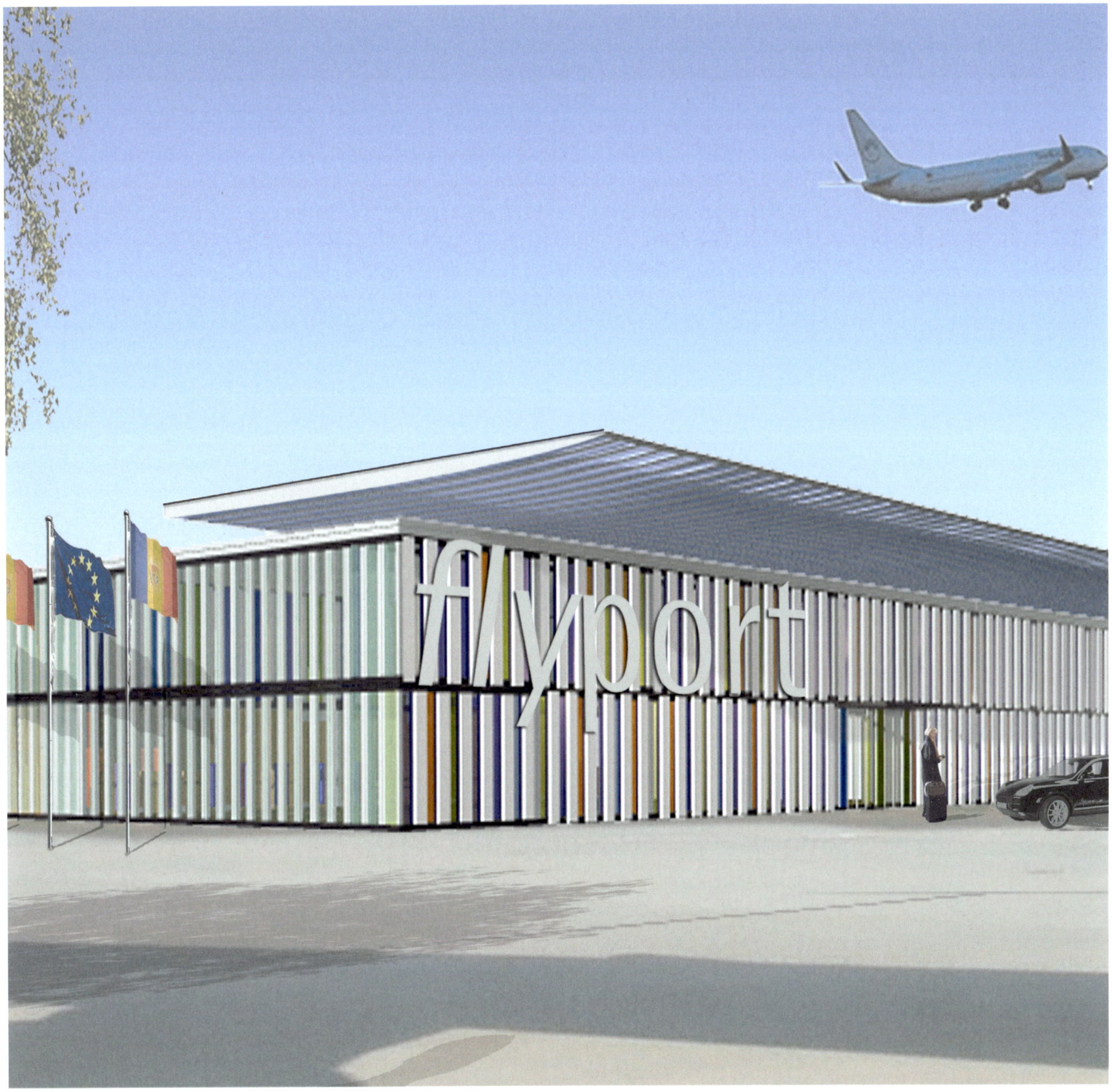

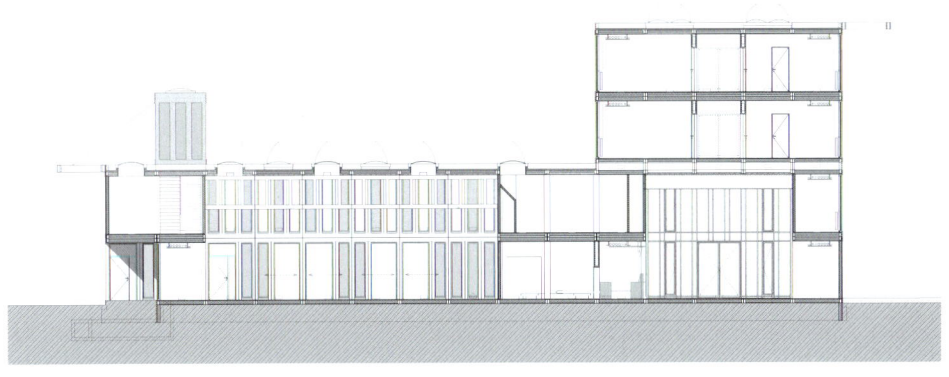

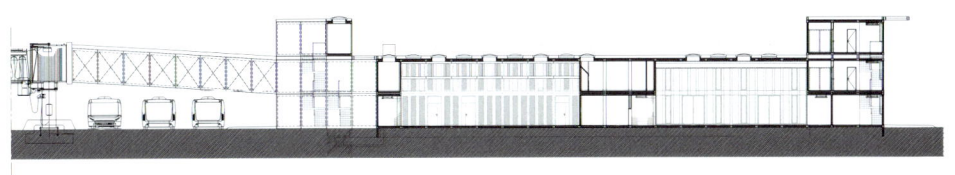

Project
IBA_DOCK
Architect
HAN SLAWIK
WITH IMS INGENIEURSGESELLSCHAFT MBH AND BOF ARCHITEKTEN

Project location | Hamburg, Germany
Estimated use | Office and exhibition space
Container type | Container frames

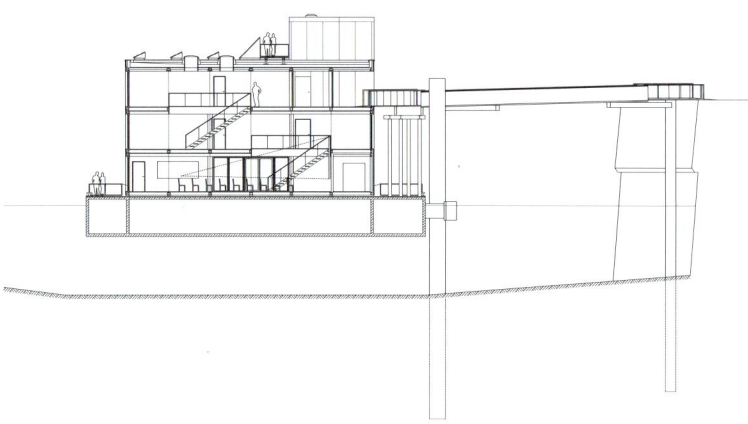

01

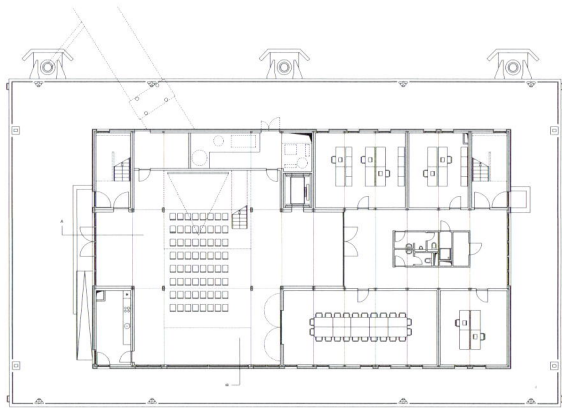

02

03

04

01, 02
Section and floor plan of the floating museum.
03
View from the dock. A bridge leads to the entrance on the top floor.
04
View from the river.
05, 06
Photos of the dock's assembly.

Germany's largest floating building bobs up and down on a grid of so-called "dolphin piles" drifting approximately 3.5 meters daily on the tide and riding atop the water through extreme storms. The building is made up of a 45×25 meter concrete pontoon that serve as jetties and create a usable floor space of 1,600 square meters over three stories. A bridge leads to the entrance on the top floor, accessing exhibition and presentation areas, a city model, cafeteria and an outdoor terrace. The modular steel frames can be disassembled for storage on land when the dock needs to pass under low bridges while being towed for maintenance work. These frames were prefabricated, transported by road, and then assembled in two weeks over the pontoon. The container modules can be easily converted at any time without having to change load bearing parts. Equipped to draw on the sun and the Elbe river for energy, the resulting building does not require any additional external power source.

Project
PIER 57
Architect
LOT-EK

Project location New York City, New York, USA
Estimated use   Event and retail spaces
Container type   Freight container

Manhattan's landmarked Pier 57 will be the site for an architecture project that will generate more than a third of a mile of new public waterfront access around its perimeter. As lead architect, local firm LOT-EK will translate the explicitly industrial building into a monumental public building, which will layer the existing four-story structure with the outdoor environment of the Hudson River Park, and its myriad cultural and leisure indoor activities. As part of the effort to make the project sustainable, the studio will reuse shipping containers to define the interiors of the pier's retail zones, a particularly apt choice in the riverside context and because so many are languishing, unused in local ports. The architects will connect the containers both vertically and horizontally by opening up their side walls, roofs and floors in various ways to create double-height as well as extra-wide interiors. The containers' ends will also be removed and glazed to allow plenty of light and views of the market to penetrate.

01, 03
The strategic clustering of the four-story containers ensures the penetration of light, air and views.
02
The plan will generate a third of a mile of new public waterfront.
04
Visitors along the waterfront esplanade.
05
A longitudinal section of the LOT-EK project.

Project
BILLBOARD BUILDING
Architect
LOT-EK

Project location   New York City, New York, USA
Estimated use   Retail space and advertisement
Container type   Freight container

02

01

03

This example of retail architecture will serve not simply as a 1,022-square-meter shop, but also as a literal billboard for the brand, while standing at the intersection of Houston Street and Bowery in Manhattan. LOT-EK, one of America's go-to architects for cargotecture, plans to use 30 freight containers to construct the space. By staggering the boxy volumes, the architects plan to create a highly textural facade with penetration of light, air and circulation that belies the typically closed nature of the shipping container.

01
Floor plans.
02
Southwest view of the retail building that will serve as its own advertising.
03
Southesast perspective of the 30-container building in Manhattan.
04
The staggered stacking of containers.

04

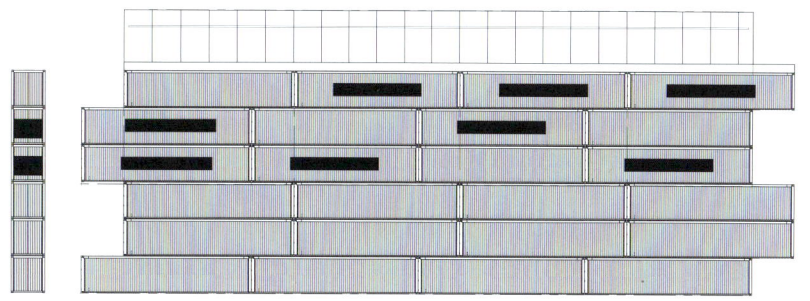

- 217 -

Project
SANLITUN SOUTH
Architect
LOT-EK

Project location Beijing, China
Estimated use   Retail and office space
Container type   Freight container

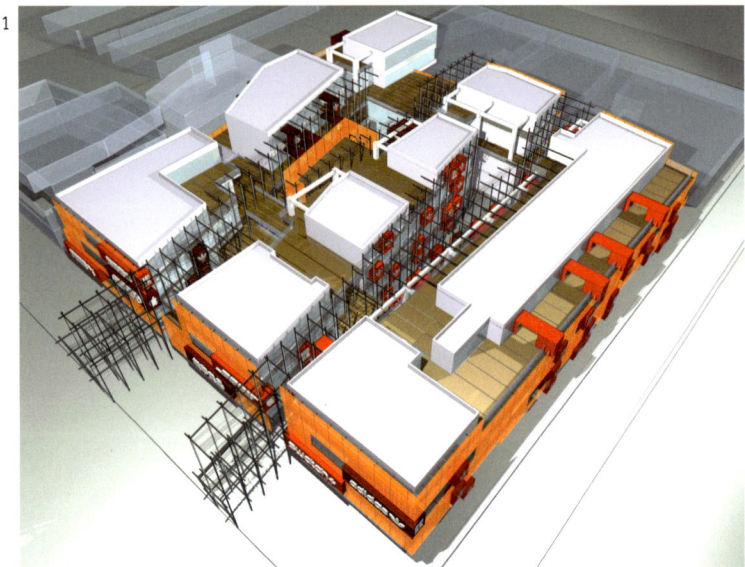

Beijing's four-story, 24,000-square-meter Sanlitun South mall is strung with shipping containers, orange-painted, stainless-steel mesh, and steel scaffolding—a form reminiscent of the medieval Chinese hutong—a quarter tightly packed with low-slung buildings threaded with alleys, elevated pathways and courtyards. New York-based LOT-EK designed the northeast section of this retail "village" to be filled with retail, restaurants and event spaces by threading it with multilevel, open-air circulation routes and 151 containers. In the alleys, the designers wedged a network of metal frames between the buildings, connecting them along the side of the buildings by a system of horizontal metal rods that serve as both railings and a brise-soleil, and creating a loggia on the upper levels. The texture of the structure derives from the standard eight-foot width of the ISO shipping containers that were plugged into the facade of the building in such a way that they stick out into the alleys and serve as canopies at the ground level: awnings on the exterior and places to display objects in shop interiors. On the higher floors, the containers are sliced through with circulation routes, forming entrances to stores or display windows and functioning everywhere as graphic objects tattooed with signage.

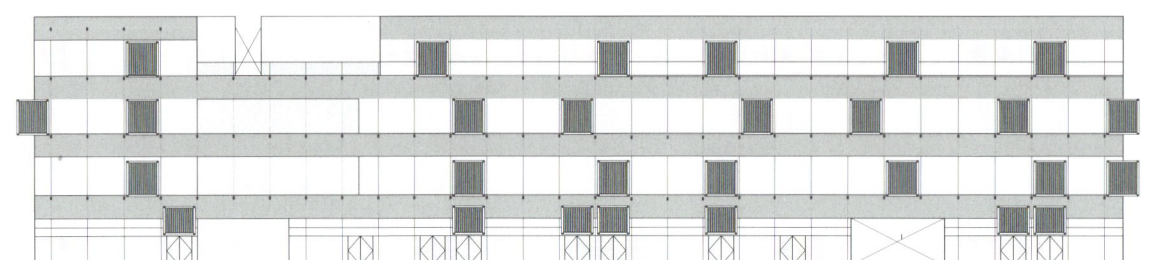

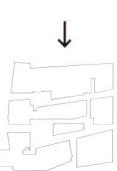

06

01, 02
The plan was based on that of the midieval Chinese hutong, a dense quarter threaded with alleys and bridges.
03, 07
Elevations.
04 – 06
Cargo containers jut out of the seemingly ordinary building forming shading outside and retail awnings within.

07

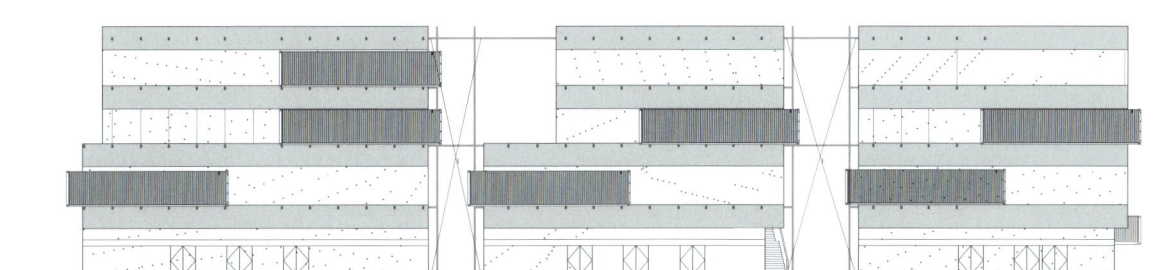
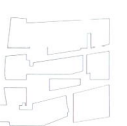

Project
RESEARCH STATION
Architect
IMS INGENIEURGESELLSCHAFT MBH AND BOF ARCHITEKTEN

Project location	Larsemann Hills, Antarctica
Estimated use	Research station
Container type	Building container

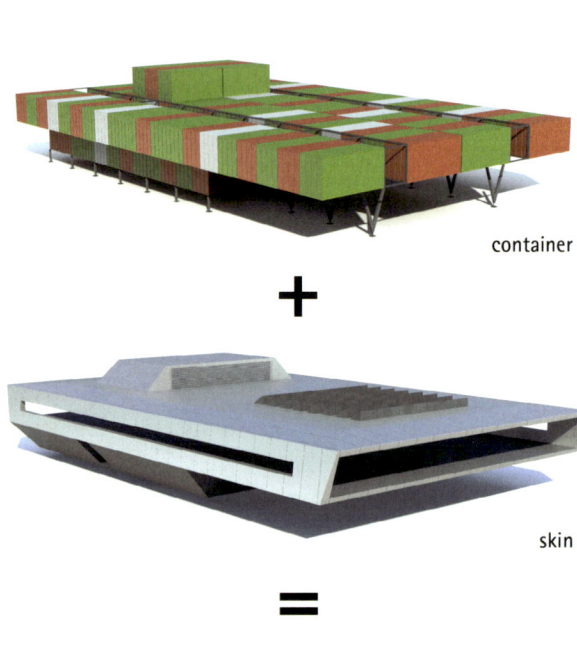

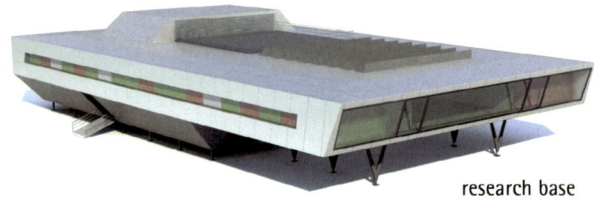

01

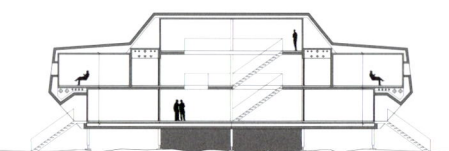

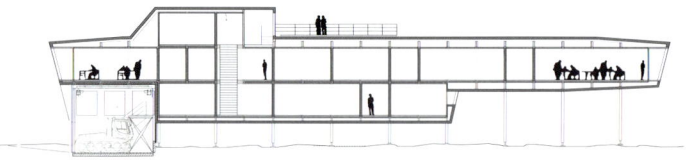

02

01
The formula for combining the station's layers: a grid of footed containers wrapped in a sleek, aerodynamic skin.
02
Longitudinal and cross section.
03
A rendering of the Antarctic station in situ. Using containers was a response to challenges of the site's extreme climate, terrain and remoteness.
04
Floor plan.

Bof Architekten were responding to extreme climate conditions and limited transport opportunities in designing this research station on a peninsula in the Larsemann Hills region of the northeastern Antarctic. Furthermore, because of the site's poor accessibility and the restrictions set forth by the Antarctic Treaty, it was crucial that the station be completely self-sufficient. The prefabricated building, which was shipped in discrete components and then assembled on site, came with its own energy supply and water preparation and treatment facilities. Assembly could only be carried out during the extremely brief Antarctic summer, but once completed, houses 25 researchers and scientists during the warm months and 15 in winter. The building consists of 132 containers, the superb mobility and flexibility of which make it well-suited to transport and short assembly times. The boxes' various programs are clearly structured and separated: The first floor features living areas with 24 single and double rooms, a kitchen and dining area, a small library, storage and washing rooms, a fitness space, offices and a lounge. Underneath, on the ground floor, the containers house laboratories, storage, equipment rooms (including the central energy unit), and a garage with a workshop. The deliberate compactness of the interiors makes for short access distances, resulting in short piping and wiring lengths. Due to low temperatures and strong winds, the containers are jacketed with an aerodynamic shell of metal panels, optimized for the prevailing climate. In terms of energy, waste heat from the CHP units is harnessed and used to heat the station, but the building also has three combined heat and power units for back-up electrical purposes. In addition, it is fitted with ventilation and air-conditioning equipment that maintain a humidity level of 30%. Windows at each end give views from the dining area and lounge onto vast expanses of ice and sea. Inside, the generous use of wood, color and ample lighting cultivate a warmer and less isolating live-work atmosphere.

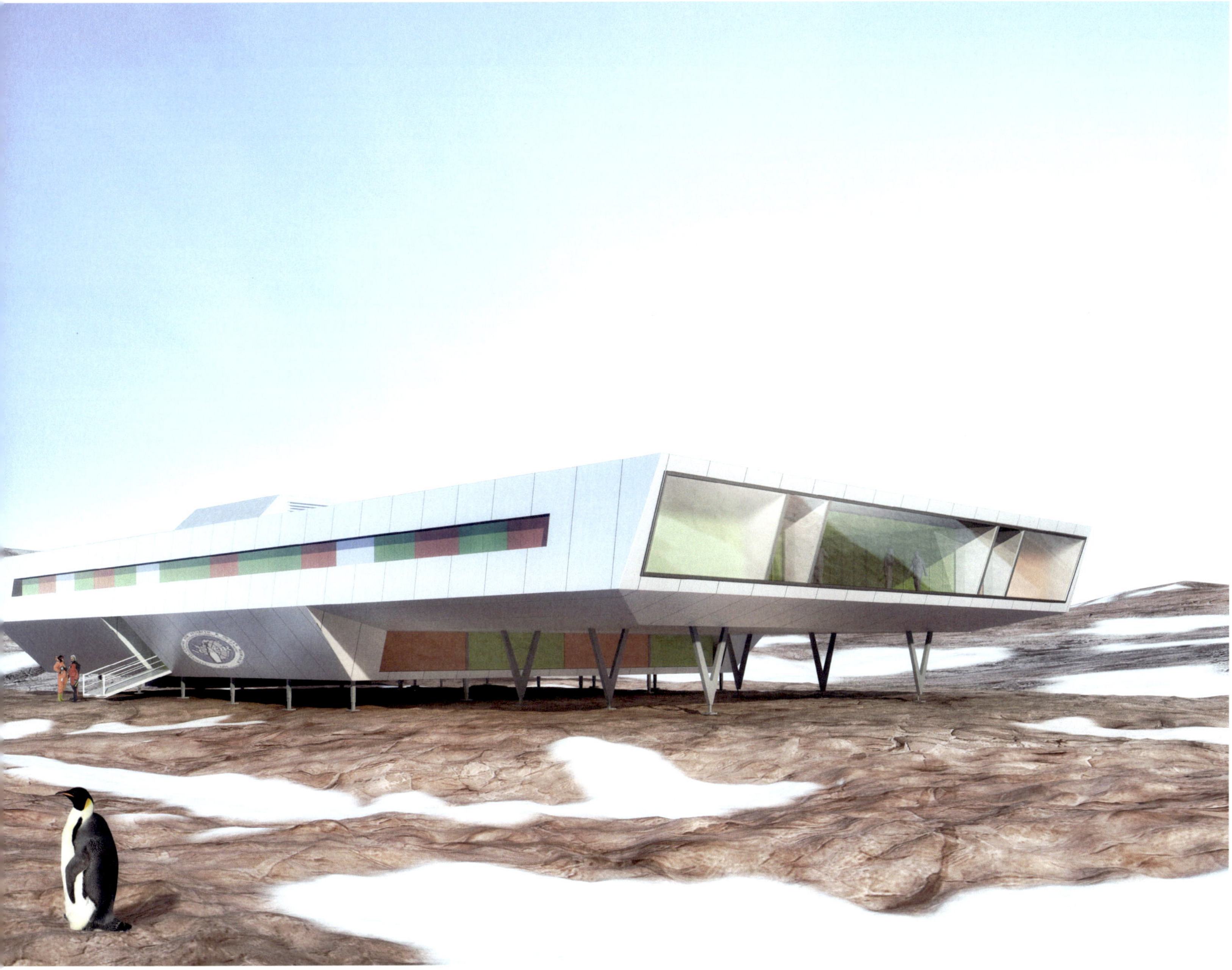

04

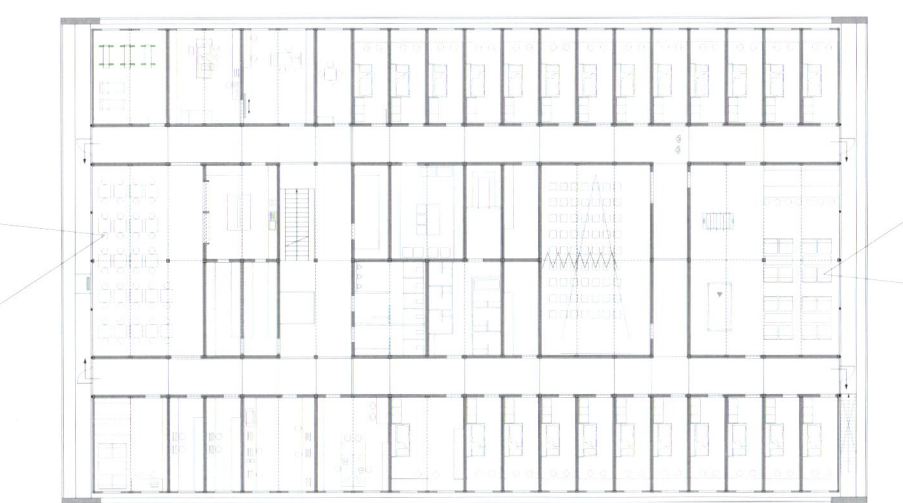

Project
SEATRAIN HOUSE

Architect
OMD – OFFICE OF MOBILE DESIGN

JENNIFER SIEGAL WITH KELLY BAIR

Project location   Los Angeles, California, USA
Estimated use   Private residence
Container type   Freight container

This 279-square-meter custom residence uses traditional, commercial and industrial materials in a playful manner. Featuring storage containers and steel found on-site in downtown Los Angeles, Office of Mobile Design created an oasis without abandoning or disguising the industrial landscape that inspired the design and provided the materials for it in the first place. Situated by the Brewery, a 300-loft live-work artist's community, the house's generous glass panels open it up, allowing natural light to pour in and connecting it visually to the rest of the community. In keeping with the creative local spirit in which the dwelling was built, the project was a collaborative experiment between the client and the fabricators who pursued a design/build approach in which creative and structural decisions were made as the house was in the process of being constructed.

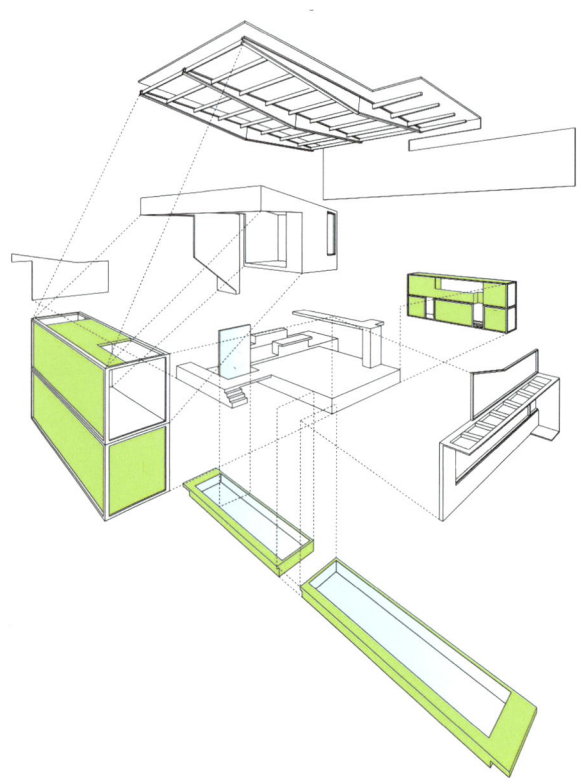

04

01, 02
Interior details and the view outside over the koi pond and lap pool.
03
The living room, an industrial sanctuary with a huge glass wall.

04
Exploded view of the building.
05 [page 223]
The house's proximity to a scrap storage site allowed the design-build team to salvage materials to preserve the site's industrial spirit.

This home literally grows up from the land around it, engaging with and incorporating the industrial history of downtown L.A. through the selection of found, on-site materials—for example, grain trailers were transformed into a koi fish pond and lap pool. Large storage containers define and separate living spaces, including entertainment, library, dining and office space (overlooking a garden below), bath and laundry, and the master bedroom, a dramatic protruding volume that wraps around the upper section of the house. This space is marked by a rich contrast of corrugated metals, industrial containers and exposed wooden beams highlighted with warm, calm green hues. All of the containers were altered in surprising ways—cut into pieces, added onto, layered or wrapped—demonstrating a plethora of possibilities in repurposing them. There are wrapped design elements throughout the house including a 12-foot-high steel plate fence that wraps the entire site, lifting up in one area to form a canopy that shades the entrance. OMD salvages and recycles materials in practical and cost-effective ways in the Seatrain project, but they also use these gracefully to create a distinctive architectural vocabulary.

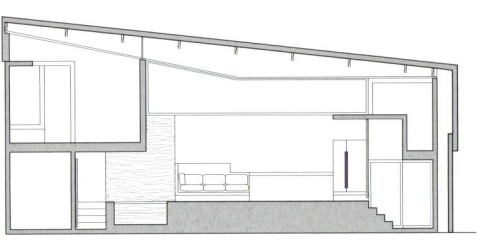

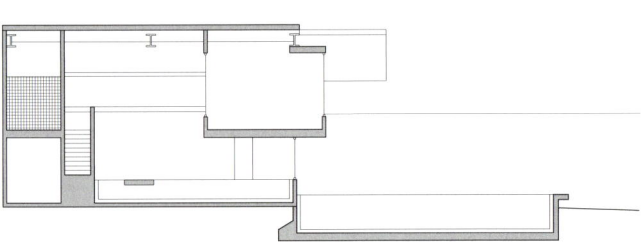

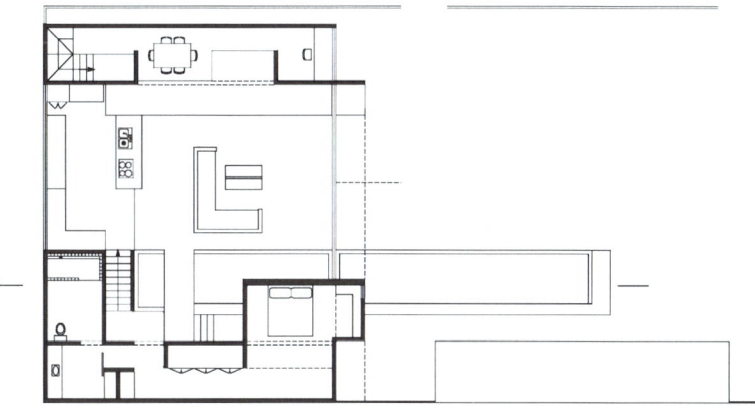

06
The home is a collage of cut-up and layered scrap materials accented with corrugated surfaces and framed in clean horizontal lines (and in counterpoint, a wild garden).
07, 10
The living room with a stone wall and corrugated metal ceiling.
08, 09
Sectional views.
11
Floor plans.

Project
ZIGLOO DOMESTIQUE
Architect
KEITH DEWEY

Project location   Victoria, British Columbia, Canada
Estimated use   Private residence
Container type   Freight container

01, 03
The front of the Zigloo residence.
02
Zigloo residence seen from the street.

Eight 20-foot cargo containers and one 10-foot container form the basis of Keith Dewey's single-family dwelling in Victoria, Canada. At 178 square meters, the interior living space is decked out with a combination of materials, including steel, glass, chain-link fence, bamboo, concrete, spray-foam insulation and wood.

Project
PARSONAGE BOYLE HEIGHTS

Architect
DEMARIA DESIGN ASSOCIATES INC.

PETER DEMARIA WITH CHRISTIAN KIENAPFEL

Project location	Los Angeles, California, USA
Estimated use	Private residence
Container type	Freight container

01
Freight containers seem to burst from the roof of this two-story dwelling.
02
View from the first floor interior.
03
The entrance.

This 390 square-meter community center and residence in East Los Angeles was the brainchild of local architect Peter DeMaria. It was built for the local church municipality by parishioners, using seven freight containers with a concrete and steel construction. Its humble but upright appearance, with horizontal lines and strong geometry, shares the beauty of Bauhaus pragmatism and was imbued by its construction crew of churchgoers with the spirit of its community.

Project
REDONDO BEACH HOUSE
Architect
DEMARIA DESIGN ASSOCIATES INC.
PETER DEMARIA WITH CHRISTIAN KIENAPFEL

Project location	Redondo Beach, California, USA
Estimated use	Private residence
Container type	Freight container

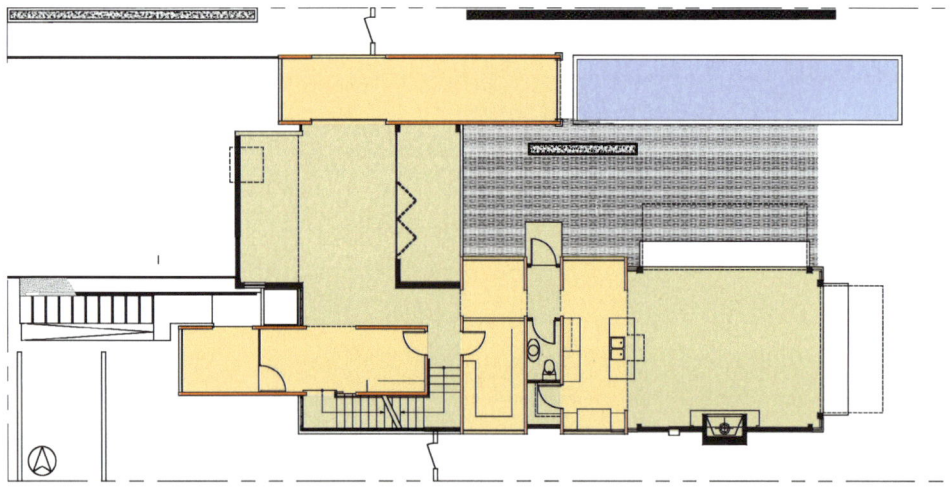

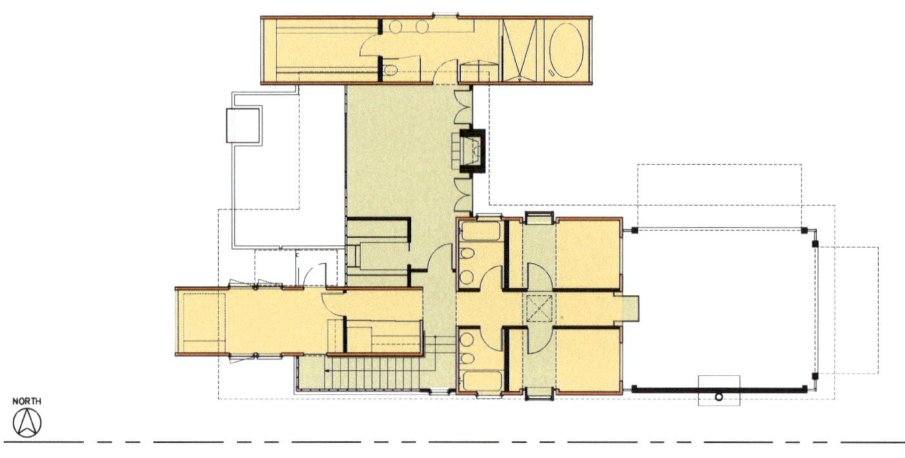

01
Ground floor plan.
02
First floor plan.
03
DeMaria used freight containers, airplane hangar doors and prefab roof panels to supplement the otherwise conventional construction of this Los Angeles area home.

Peter DeMaria built the Redondo Beach House from ISO cargo containers, amongst other materials, thus combining industrial materials usually not used for residential construction with materials employed by conventional building methods. The result is what he calls a "hybrid". Old containers, hangar doors, prefabricated roof panels, greenhouse multi-skinned acrylic sheets, formaldehyde-free plywood, sprayed insulating material, and tankless water heaters were used. Using containers appealed to the architect because of their robustness and because they are virtually mold and termite-proof. In his design, DeMaria drew on tried and tested mass products: from Frank Lloyd Wright's Textile Block homes and Andy Warhol's prints, all the way to McDonald's mass-production of hamburgers. His aim was bring more people closer to their dream of owning a high-quality custom home at an affordable price. "This project is the torchbearer for a new, more affordable method of design and construction," says DeMaria, "Architecture as a Product."

04

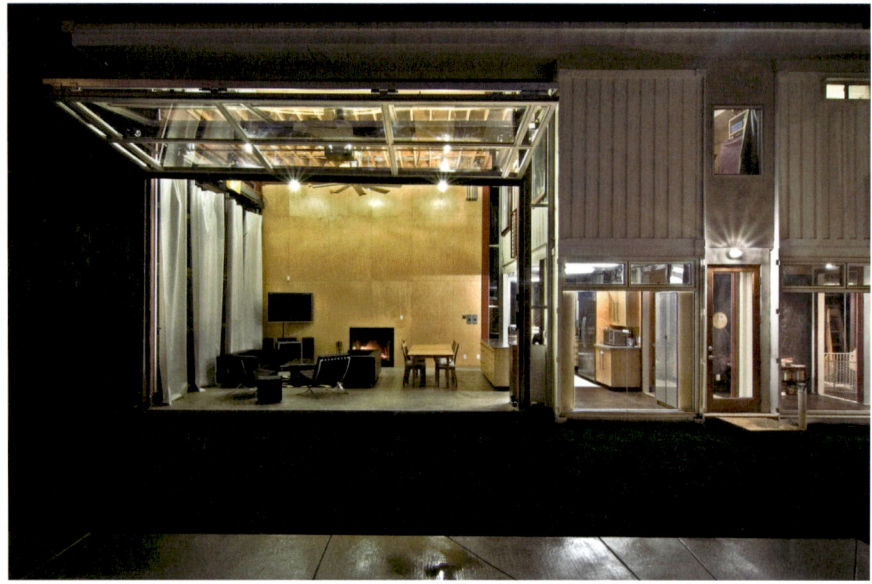

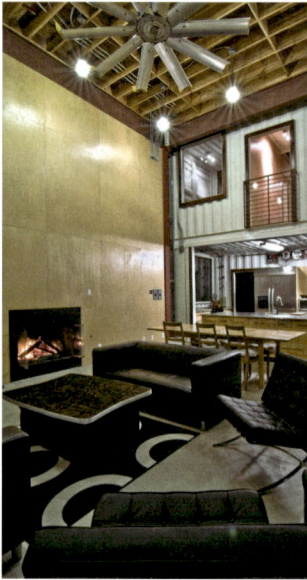

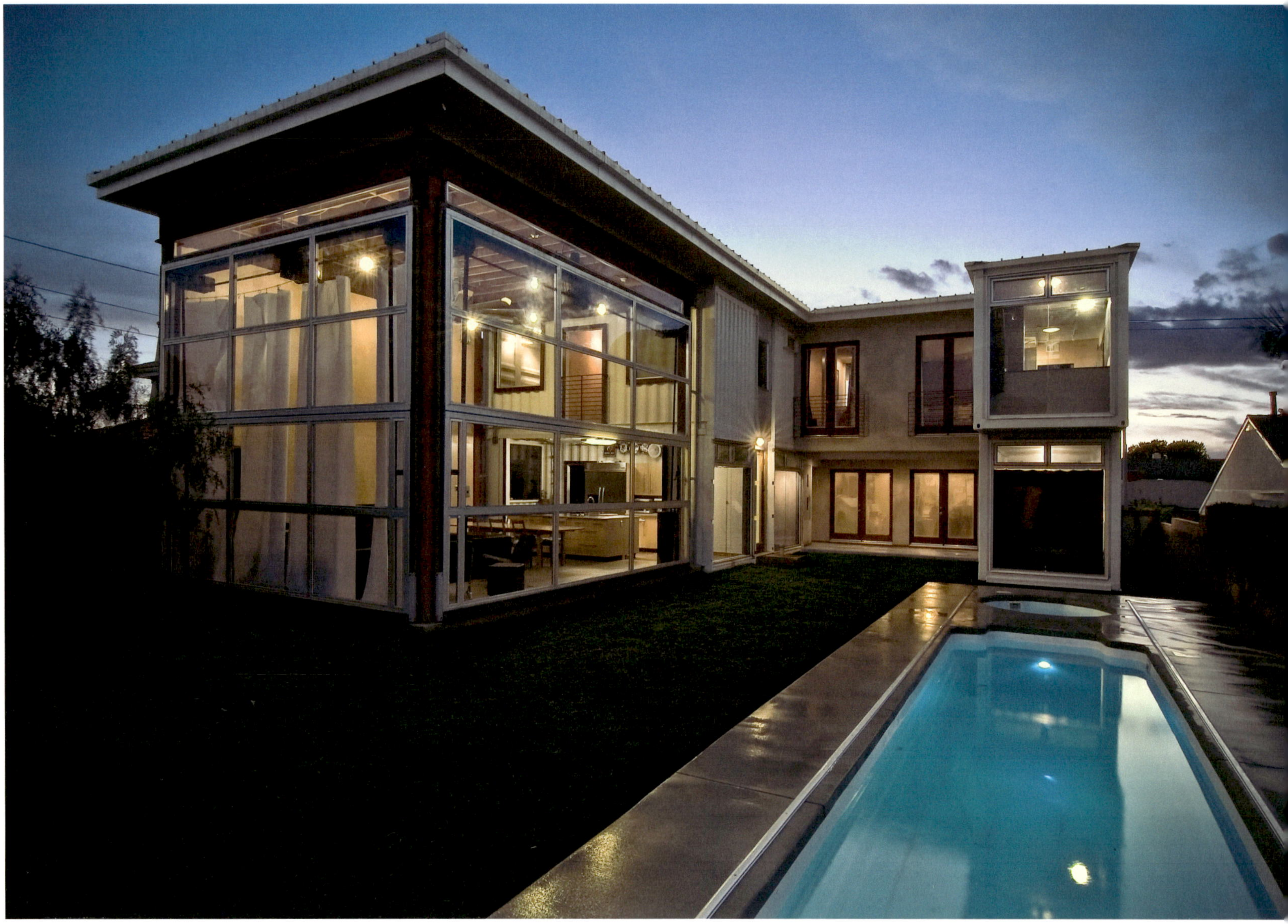

05

06

04
The backyard with swimming pool and large window surfaces.
05
The windows swing upward to combine living room with lawn.
06
Interior details.
07
The entrance to the house.

Project
CHALET DU CHEMIN BROCHU
Architect
PIERRE MORENCY ARCHITECTE

Project location   Beaulac-Garthby, Québec, Canada
Estimated use   Private residence
Container type   Freight container

01

02

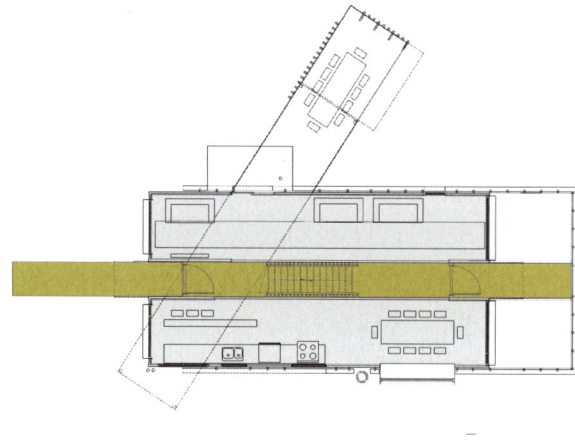

03

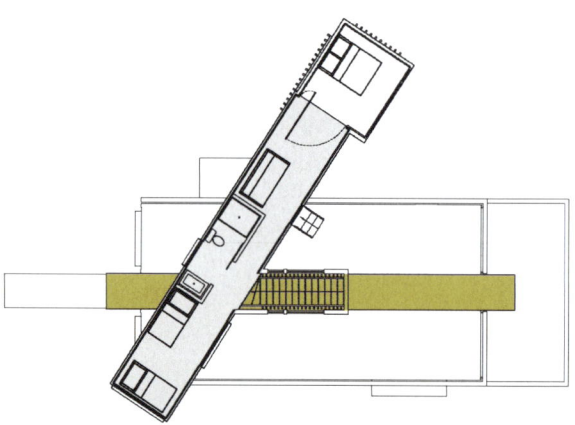

04

 The chalet's prefabrication began in Montreal with the acquisition of three used black containers, which were then transported to the site and suspended like immense beams amongst the trees. A second, more traditional wooden skin, made from local hemlock that was dried and hammer-marked with the help of a portable sawmill, was applied to join the metal boxes together.

05

06

07

08

09

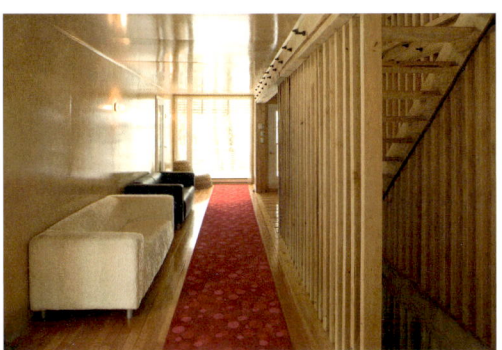

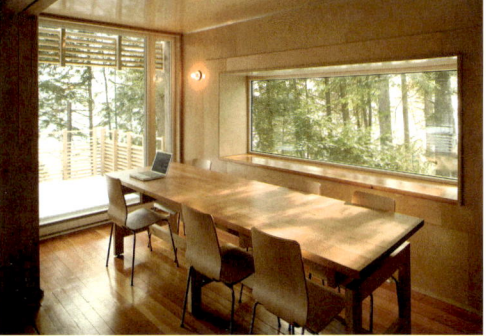

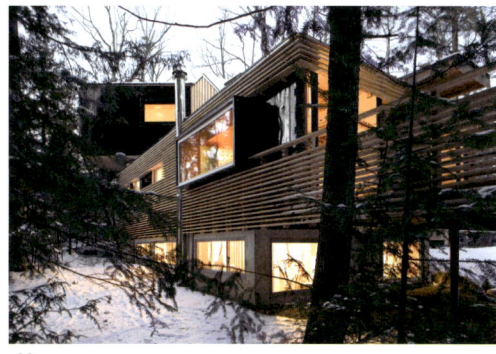

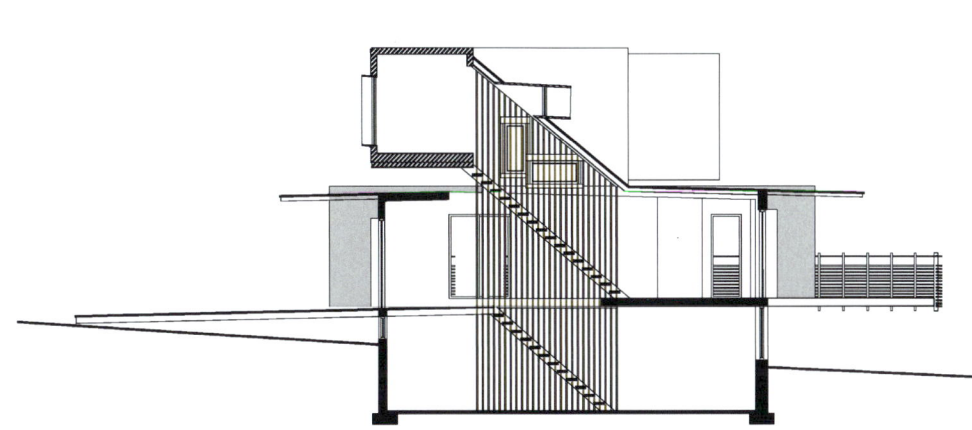

01, 02
Three black cargo containers form the core of this second home but they are clad in a second skin made from strips of local hemlock wood.
03
Ground floor plan.
04
First floor plan.
05, 08
The rear deck is where the containers are most visible. Morency let them jut out slightly, emphasising their elegant placement, which makes the most of forest views.
06, 07
Interiors.
09
Longitudinal section.

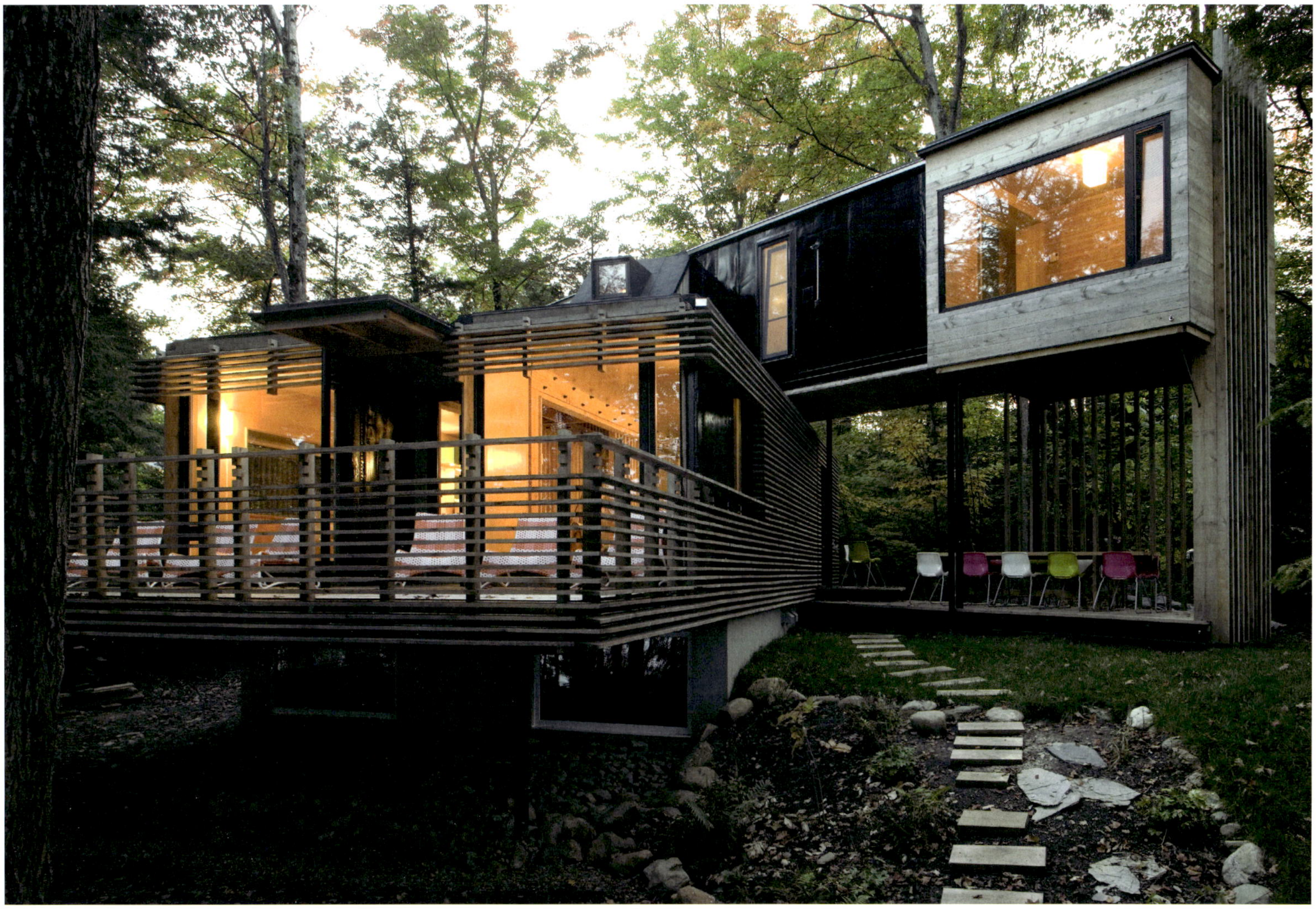

Project
HIDDEN VALLEY
Architect
MARMOL RADZINER PREFAB

Project location   Moab Desert, Utah, USA
Estimated use   Private residence
Container type   Container frames

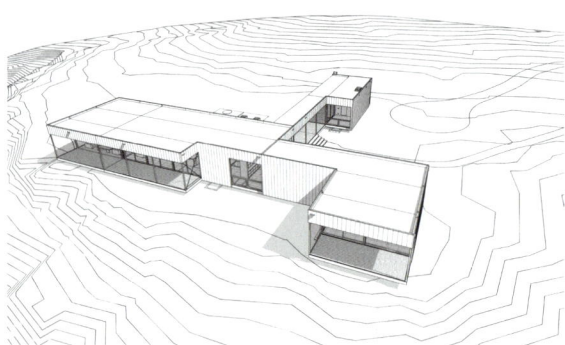

Los Angeles-based architects Marmol Radziner designed their Hidden Valley prefabricated house as a single-family vacation home near Moab, Utah. The firm's careful siting, emphasis on overlapping indoor and outdoor living, and integration of sustainable design elements celebrate and make the most of the area's inimitable access to nature. The 230-square-meter dwelling includes a 160-square-meter terrace and sits on an open, hundred-acre site punctuated by red rock formations and desert cliffs. Its two-bedroom, two-bath structure consists of five interior modules, seven ipe-wood deck modules and three additional modules making up the garage, all of which were fabricated in a factory and shipped fully assembled. In the factory, precut steel beams and joists were welded together to create the floor and roof frames. Pre-cut columns were attached to the floor frame and the roof structure was placed atop the columns. Then workers installed structurally insulated panels (SIPs) to form the sub-floor and roof structures. Once the steel frame was set, the interior wall framing, plumbing, and electrical and mechanical components were installed. Finally, they added windows and doors, interior and exterior finishes, built-in casework, appliances, and fixtures. The modules were then transported to the site on flatbed trucks, lifted onto the foundation with a crane, bolted together and welded to the foundation to create a permanent structure.

01 – 03
The prefab house peers over a desert cliff in Utah.
04
Site plan: Careful situating of the house allowed the team to integrate indoor and outdoor living.
05
The two-bedroom house consists of 15 factory-built modules.
06
A living room-cum-deck with pool.
07
Floor plan.

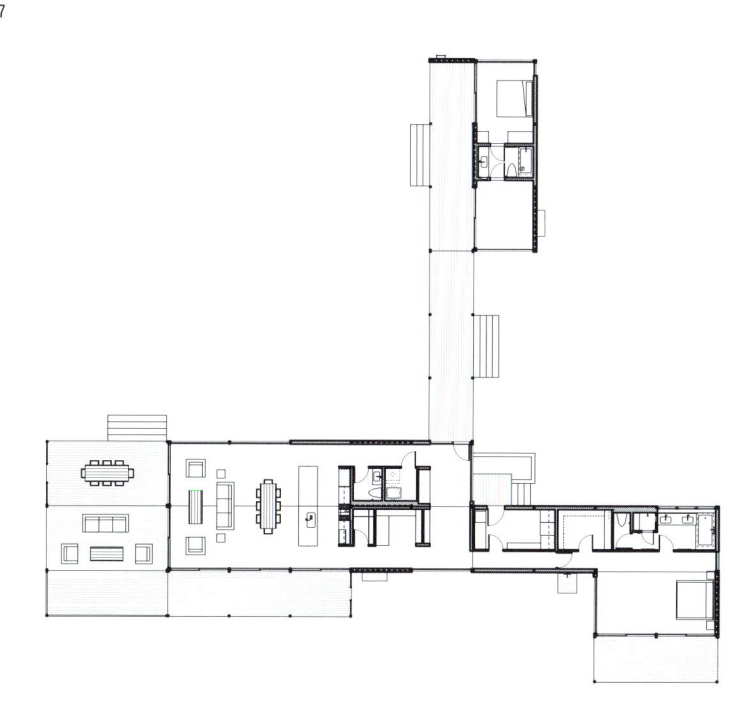

The Hidden Valley residence features sustainable materials and systems to minimize the environmental impact of both its production and its use: geothermal systems use the earth to heat and cool the home, while solar panels draw energy from the sun. The home's recycled steel frame promises long-term durability against the elements, termites and even mold without requiring chemical treatments.

08
Kitchen and living room. The house draws on geothermal energy and solar panels.
09
A bedroom with views of red rock and blue sky. The reverence for nature isn't just expressed through views but through sustainable systems that power the dwelling.
10
In many ways, the sturdy recycled steel architectural frame becomes a picture frame in the dramatic landscape.

08

OUTLOOK

With its numerous and varied architectural examples, this container atlas demonstrates quite clearly that containers are complex building modules that offer suitable solutions for a wide range of uses. The wide spectrum of architectural solutions involving containers, some of which are quite experimental, demonstrate a high und impressively consistent level of quality. The way in which containers are used as building modules appears to be opening up the possibility of a creative, sophisticated, almost playful manner of working with modular space. Unexpected solutions are being realized in unconventional ways and are attracting a lot of attention. The negative image of container buildings, which arose over the last decades due to temporary, low-quality structures, has long since been reversed. Construction based on containers is now "in". Spatial solutions that exploit the potential of these globalized vessels can now boast a progressive and intelligent image.

However, the development of these intelligent spatial solutions with (freight) containers or other related building systems definitely requires architectural input and expertise. For this to happen, background knowledge about the potential possibilities and also the limitations of container architecture is necessary; the potential and the limitations are also influenced by engineering constraints. The present compendium provides this knowledge that, because of the pace of developments in container architecture, actually needs to be continually updated. Every month, new and interesting container projects are being implemented; nonetheless, the fundamentals of working with containers remain universally valid.

How will the freight container sector develop in the future? Will there be developments that influence the use of containers in architecture? There is no doubt that international freight transportation is no longer conceivable without containerization. The integrated logistical tools used in shipping will not change again in a revolutionary way in the foreseeable future. The continued availability of freight containers as a building block is thus assured. Nonetheless, there are indications that the dimensions of containers, and thus also of transportation equipment, may need to be improved. Some changes have already been carried out in the past—for example, the introduction of the high cube format that was one foot (approx. 30 centimeters) higher, the isolated introduction of 45-foot-long containers, and the even less common widening of containers by a few centimeters to achieve the maximum of 2.5 meters allowed by German road traffic regulations. There are also other conceivable ways of further optimizing container transportation—e.g. the widespread use of flexible container types with folding, sliding, or flap mechanisms could help to save valuable space on return journeys. All these developments have influenced and will continue to influence the use of containers in construction.

Container architecture has great potential with regard to sustainability. When viewed from an ecological perspective, container building blocks offer a lot of features that fulfill contemporary ideas regarding the absolute flexibility of spatial solutions: they are a variable and mobile minimal building block for extensions to existing buildings and for new projects, providing quick, inexpensive spatial solutions that can be converted, reused or recycled in a sustainable manner later on. "Container Architecture" thus has major potential for delivering adaptable building solutions that will remain relevant even when requirements change in the future.

൦9

APPENDIX

Timeline
Index of Images
Bibliography
List of companies
Organizations
Regulatory works
Glossary
Index of projects
Editors

Appendix

TIMELINE

1990

HOORN BRIDGE
LUC DELEU
Page 134

1992

CAMPUS
HAN SLAWIK
Page 184

1997

SPACEMAN SPIFF
SERDA ARCHITECTS
Page 84

2000

HOLYOKE CABIN
PAUL STANKEY, SARAH NORDBY
Page 98

TEMPORARY ACCOMMODATION FOR 1001 DAYS
MEYER EN VAN SCHOOTEN
Page 190

2001

FUTURE SHACK
SEAN GODSELL
Page 102

TRINITY BUOY WHARF (CONTAINER CITY I)
NICHOLAS LACEY AND PARTNERS
Page 168

2002

12 CONTAINER HOUSE
ADAM KALKIN
Page 148

50,000€ MODULAR-HOUSE
MM+P / MEYER-MIETHKE & PARTNER
Page 189

BED BY NIGHT
HAN SLAWIK
Page 162

BOHEN FOUNDATION
LOT-EK
Page 54

MYSTERY CUBE
HAN SLAWIK
Page 186

PLATOON CULTURAL DEVELOPMENT BERLIN
PLATOON.BERLIN & SOEREN ROEHRS
Page 116

SANITARY FACILITIES FOR SUMMER CAMP
AFF ARCHITEKTEN
Page 94

TRINITY BUOY WHARF (CONTAINER CITY II)
NICHOLAS LACEY AND PARTNERS
Page 168

2003

CONSTRUCTION X
LUC DELEU
Page 134

MDU / MOBILE DWELLING UNIT
LOT-EK
Page 53

RAINES COURT
ALLFORD HALL MONAGHAN MORRIS
Page 174

SEATRAIN HOUSE
OMD – OFFICE OF MOBILE DESIGN
Page 222

SPEYBANK
LUC DELEU
Page 134

THEATERHAUS STUTTGART
WITH ENGELHARDT.EGGLER.ARCHITEKTEN.
Page 74

WISMAR TECHNOLOGY AND RESEARCH CENTER
JEAN NOUVEL WITH ZIEBELL & PARTNER
Page 180

2004

B-CAMP
HELEN & HARD ARCHITEKTEN
Page 156

BOOK FAIR PAVILION
FRADE ARQUITECTOS
Page 153

C320S
HYBRID, CARGOTECTURE
Page 96

CRUISE CENTER
RENNER HAINKE WIRTH ARCHITEKTEN
Page 208

KING KAMEHAMEHA BEACH CLUB
ANGELA FRITSCH ARCHITEKTEN
Page 152

ORBINO
LUC DELEU
Page 132

SEVEN-DAY ARCHITECTURE
LHVH ARCHITEKTEN
Page 188

SUBURBAN HOUSE KIT
ADAM KALKIN
Page 144

UNIQLO FLAGSHIP
LOT-EK
Page 66

2005

CANCER CENTER AMSTERDAM
MVRDV
Page 178

CONTAINING LIGHT MOBILE UNIT
EER ARCHITECTURAL DESIGN
Page 60

CONTAINER HOUSING
GUSTAU GILI GALFETTI
Page 64

GOLD PAVILION
ANGELA FRITSCH ARCHITEKTEN
Page 95

INTERSTITIAL LIVING
LHVH ARCHITEKTEN
Page 115

MIRROR ERROR
YASUTAKA YOSHIMURA
WITH MANABU MIZUNO
Page 90

NOMADIC MUSEUM
SHIGERU BAN ARCHITECTS
Page 200

PUSH BUTTON HOUSE
ADAM KALKIN
Page 57

QUBIC AMSTERDAM
HVDN ARCHITECTEN
Page 176

SAUNA BOX
CASTOR DESIGN
Page 100

WIJN OF WATER
BIJVOET ARCHITECTUUR & STADSONTWERP
Page 108

2006

ARCHITEKTURBOX
SHE-ARCHITEKTEN
Page 160

BORNACK DROP STOP TRAINING CENTER
PATZNER ARCHITEKTEN
Page 70

CONTAINER HOUSE (KILLER HOUSE)
ROSS STEVENS
Page 118

FREITAG FLAGSHIP STORE ZURICH
SPILLMANN ECHSLE ARCHITEKTEN
Page 122

Appendix

GAD
MMW ARCHITECTS OF NORWAY
Page 138

HALF MILE
LEIBNIZ UNIVERSITY HANOVER
Page 127

KEETWONEN
TEMPOHOUSING / JMW ARCHITEKTEN
Page 172

KIOWA PROTOTYPE (HABITAINER)
LUIS RODRÍGUEZ ALONSO, JAVIER PRESA
Page 52

MEETINGPOINT PLAN06
LHVH ARCHITEKTEN
Page 126

PAPERTAINER MUSEUM
SHIGERU BAN ARCHITECTS +
KACI INTERNATIONAL INC.
Page 204

UNIQLO POP-UP STORES
LOT-EK
Page 66

VOLVO C30 EXPERIENCE PAVILION
KNOCK
Page 198

ZIGLOO DOMESTIQUE
KEITH DEWEY
Page 226

2007

LIVE/WORK SPACE
SCULP(IT) ARCHITECTEN
Page 124

CHALET DU CHEMIN BROCHU
PIERRE MORENCY ARCHITECTE
Page 232

CHILDREN'S ACTIVITY CENTRE
PHOOEY ARCHITECTS
Page 110

DERCUBE
LHVH ARCHITEKTEN
Page 187

HIDDEN VALLEY
MARMOL RADZINER PREFAB
Page 234

ILLY CAFÉ
ADAM KALKIN
Page 56

KOELNERBOX (PROTOTYPE)
JAN HOHLFELD ET AL.
Page 88

PORT-A-BACH
ATELIERWORKSHOP
Page 50

R4HOUSE
LUIS DE GARRIDO
Page 164

SJAKKET YOUTH CENTER
PLOT = JDS+BIG
Page 81

SKAEVE HUSE
TEMPOHOUSING / KERSSEN LIJBERS
Page 69

X-SPACE ATELIER
FPS OFICINA DE ARQUITECTURA
Page 80

2008

CONTAINERART SÃO PAULO
ARTUR LESCHER, BERNARDES+JACOBSEN
Page 136

DINAHOSTING OFFICES
O ANTIDOTO
Page 62

EICHBAUMOPER
RAUMLABORBERLIN
Page 130

FANSHOP OF GLOBALIZATION
RAUMTAKTIK
Page 89

GORMAN SHIP SHOP
NEST ARCHITECTS
Page 91

OLD LADY HOUSE
ADAM KALKIN
Page 145

PUMA CITY
LOT-EK
Page 140

REDONDO BEACH HOUSE
DEMARIA DESIGN ASSOCIATES INC.
Page 228

SANLITUN SOUTH
LOT-EK
Page 218

SKY IS THE LIMIT
BUREAU DES MÉSARCHITECTURES
Page 76

2009

BRENTWOOD CABANA
IC GREEN INC.
Page 68

CONSUMER TEMPLE – BROKEN ICON
GARY DEIRMENDJIAN
Page 135

CONTAINR CINEMA
ROBERT DUKE ARCHITECT ET AL.
Page 128

HOMEBOX1
HAN SLAWIK
Page 92

INFINISKI MANIFESTO HOUSE
JAMES & MAU ARCHITECTS
Page 104

PARSONAGE BOYLE HEIGHTS
DEMARIA DESIGN ASSOCIATES INC.
Page 227

PLATOON KUNSTHALLE SEOUL
PLATOON AND GRAFT ARCHITECTS
WITH U-IL ARCHITECTS
Page 192

RESIDENTIAL CONTAINERS
HSH ARCHITECTS
Page 78

2010

IBA_DOCK
HAN SLAWIK
WITH IMS INGENIEURGESELLSCHAFT MBH
AND BOF ARCHITEKTEN
Page 214

IN PROGRESS

BILLBOARD BUILDING
LOT-EK
Page 217

PIER 57
LOT-EK
Page 216

RESEARCH STATION
IMS INGENIEURGESELLSCHAFT MBH
AND BOF ARCHITEKTEN
Page 220

TEMPORARY EXHIBITION / 100 YEARS OF FOOTBALL
CLUB ST. PAULI
KOMMA4 ARCHITEKTEN, PÜTZ / REETZ
Page 114

PLANNED

FLOATING HOUSE
HAN SLAWIK
Page 86

FLYPORT
WOLFGANG LATZEL ARCHITEKTEN
Page 212

INDEX OF IMAGES

Image 001 – 004: Slawik et al. (Eds.)
Image 005: Latzel Architekten
Image 006: Photographer: Christian v. Steffelin
Image 007: Photographer: Andre Movsesyan
Image 008: Photographer: Peter Bennetts
Image 009: Photographer: Danny Bright
Image 010: Photographer: Michael Moran Photography
Image 011: Photographer: Luca Campigotto
Image 012: Photographer: Philippe de Gobert
Image 013: Photographer: Felix Luong
Image 014: Photographer: Philippe Ruault
Image 015: Slawik
Image 016: Mecanoo Architekten
Image 017 – 024: Slawik et al. (Eds.)
Image 025: Slawik
Image 026 – 028 Slawik et al. (Eds.)
Image 029: Photographer: Philippe de Gobert
Image 030: Slawik

Image 031: Photographer: Luuk Kramer
Image 032, 033: Slawik et al. (Eds.)
Image 034: conform ISO
Image 035, 036: Buchmeier
Image 037: Slawik et al. (Eds.)
Image 038: nach ISO
Image 039: Slawik et al. (Eds.)
Image 040 – 046: Buchmeier
Image 047: Hafen Hamburg
Image 048: Slawik et al. (Eds.)
Image 049: Singamas Ltd.
Image 050, 051: Buchmeier
Image 052 – 056: Slawik et al. (Eds.)
Image 057: Portpictures.nl
Image 058 – 059: Slawik et al. (Eds.)
Image 060, 061: Tinney, KLEUSBERG GmbH & Co. KG
Image 062: Slawik
Image 063: Photographer: ALHO Systembau GmbH / Herr Schade
Image 064: Photographer: Axel Ollenschläger
Image 065: bof architekten
Image 066: Slawik et al. (Eds.)
Image 067: Tinney, KLEUSBERG GmbH & Co.KG
Image 068: ALHO Systembau GmbH
Image 069: Tinney, KLEUSBERG GmbH & Co.KG
Image 070, 071: Slawik et al. (Eds.)
Image 072, 073: ALHO Systembau GmbH
Image 074 – 079: Tinney, KLEUSBERG GmbH & Co.KG
Image 080 – 082: ALHO Systembau GmbH
Image 083: Slawik
Image 084: Slawik et al. (Eds.)
Image 085: Photographers: Slawik, Frank Aussieker
Image 086 – 088: Slawik et al. (Eds.)
Image 089 – 091: ALHO Systembau GmbH
Image 092 – 096: Slawik

Appendix

BIBLIOGRAPHY

CONTAINERS IN GENERAL

Container Contacts – 2009 Edition (mit CD-ROM). 37. Auflage, Storck Verlag, Hamburg, 2009.

Kotnik, Jure: *Container Architecture: This Book Contains 6,441 Container.* Links books, Barcelona, 2008.

Smith, Courtenay / Topham, Sean: *Xtreme houses.* Prestel, München, 2002.

Siegal, Jennifer: *MOBILE: The Art of Portable Architecture.* Princeton Architectural Press, 2002.

Siegal, Jennifer: *More mobile: Portable Architecture for Today.* Princeton Architectural Press, 2008.

Topham, Sean: *Move House.* Prestel, München, 2004.

FREIGHT CONTAINERS

Beplat, Klaus: *Probleme der Einführung des Containerverkehrs.* Verlag Weltarchiv, Hamburg 1970.

Cudahy, Brian J.: *Box Boats: How Container Ships Changed the World.* Fordham Univ. Press, New York, 2006.

Eilers, Reimer: *Das neue Tor zur Welt: Vierzig Jahre Container im Hamburger Hafen.* mareverlag GmbH, Hamburg, 2009.

Gräfing, Birte / Heinrichs, Dirk (Hrsg): *Vom Stauhaken zum Container: Hafenarbeit im Wandel.* Kellner, Bremen, 2008.

International Organization for Standardization (Hrsg.): *Freight containers: ISO Standards Handbook.* ISO Central Secretariat, Genf, 2000.

Kalkin, Adam: *Architecture and Hygiene.* Batsford, London, 2002.

Kalkin, Adam: *Quik Build: Adam Kalkin's ABC of Container Architecture.* Bibliotheque McLean, London 2008.

Kienzle, Jörg: *Das Container-Transportsystem.* In: Schriftenreihe Transport und Verkehr 1, München, 1991.

Lamster, Mark / Tolla, Ada: *Urban Scan / LOT-EK.* Princeton Architectural Press, New York, 2002.

Levinson, Marc: *The Box: How the Shipping Container Made the World Smaller and the World Economy Bigger.* Princeton University Press, Princeton NJ, 2006.

Lindner, Erik: *Die Herren der Container: Deutschlands Reeder-Elite.* Hoffmann und Campe, Hamburg, 2008.

Pawlik, Thomas / Hecht, Heinrich: *Containerseeschifffahrt.* Heel Verlag GmbH, Königswinter, 2007.

Preuß, Olaf: *Eine Kiste erobert die Welt: Der Siegeszug einer einfachen Erfindung.* Murmann Verlag, Hamburg, 2007.

Scoates, Christopher / Neidhardt, Jane: *LOT-EK: Mobile Dwelling Unit.* D.A.P. Distributed Art Publishers, New York, 2003.

Sawyers, Paul: *Intermodal Shipping Container Small Steel Buildings.* 2005, 2008.

Smith, John: *Shipping Containers as Building Components, Resarch.* NA Brighton University, UK, School of the Built Environment, 2006.

Witthöft, Hans Jürgen: *Container: Eine Kiste macht Revolution.* Koehler, Hamburg, 2000.

Witthöft, Hans J.: *Container: Die Mega-Carrier kommen.* Koehler, Hamburg, 2004.

BUILDING CONTAINERS

Doßmann, Axel / Wenzel, Jan / Wenzel, Kai: *Architektur auf Zeit: Baracken, Pavillions, Container.* b_books Verlag, Berlin, 2006.

Uhde, Robert: *De Vijf: Student Housing in Utrecht, Netherlands – Living in a box.* In: Architektur aktuell, 5/2005, S. 86-91, Springer Verlag, Vienna.

CONTAINER FRAMES

Luig, Klaus Th. / Lenze, Veronika: *Stahl im Wohnungsbau.* Ernst & Sohn Verlag, Berlin, 1998.

Staib, Gerald / Dörrhöfer, Andreas / Rosenthal, Markus: *Elemente und Systeme: modulares Bauen.* Birkhäuser, Edition DETAIL, Basel, 2008.

Stahl Informationszentrum: *Dokumentation 548, Kostengünstiger Wohnungsbau mit Stahl.* Symposium, Düsseldorf, September 22, 1998.

LIST OF COMPANIES

FREIGHT CONTAINERS

MANUFACTURERS

www.bslcontainers.com
BSL Container Manufacturer, Hongkong

www.containerparts.nl
Van Doorn Container Parts B.V., Rotterdam/
the Netherlands

www.dcmhl.com
DCM Hyundai Limited, Faridabad/India

www.hempel.com
Hempel A/S, Lyngby/Denmark;

www.korindo.co.id
Korindo Group, Jakarta/Indonesia

www.sec-bremen.de
Ship's Equipment Centre, Bremen/Germany

www.singamas.com
Singamas Container Holdings Ltd., Shanghai/China

www.sino-peak.com
Sino-Peak Container Manufacturing Co. Ltd.,
Tianjin/China

DEALERS

www.alcu.de
Container und mobile Räume GmbH,
Seevetal/Germany

www.conical.de
CONICAL GmbH, Hamburg/Germany

www.containex.com
CONTAINEX Container — Handelsgesellschaft mbH,
Wiener Neudorf/Austria

www.csshippingcontainers.co.uk
CS Chipping Containers, Battisford/GB

www.finsterwalder-container.de
Finsterwalder Container GmbH,
Kaufbeuren/Germany

www.magellan-maritime.de
Magellan Maritime Services GmbH,
Hamburg/Germany

www.rainbow-containers.de
Rainbow Containers GmbH, Biedersdorf/Germany

www.renz-container.com
RENZ Handel & Logistik GmbH, Stuttgart/Germany

BUILDING CONTAINERS

www.algeco.de
www.alho.com
www.ass-container.de
www.bboxx.de
www.buero-wohncontainer.de
www.chs-container.de
www.container.de
www.containerbau.de
www.containerland.de
www.containersachverstand.de
www.containervermietung.de
www.containerwelt.de
www.csraum.de
www.deutsche-industriebau.de
www.drehtainer.de
www.easy-container.de
www.eberhardt.eu
www.graeff-gmbh.de
www.kleusberg.de
www.militaer-container.de
www.modulplan.de
www.mvs-zeppelin.de
www.optirent-mietservice.de
www.panzerbaer.de
www.pilz-container.de
www.raumsysteme.de
www.saebu.de
www.sbs-containerservice.de
www.siko-container.de
www.acm-container.com
www.flexotel.nl
www.portacabin.co.uk
www.containex.com
www.chv.at
www.a1container.at
www.uniteamcontainer.com
www.ofc.cz
www.uniteam.se
www.prqu.com
www.touax.es
www.grupoprasur.com
www.officetrailer.com
www.actonmobile.com
www.modspace.com
www.b2b-exchange.com
www.prefabcomparison.com
www.office-trailer.com (trailers)
www.swmobilestorage.com
www.360mobileoffice.com
www.wilmotmodular.com
www.triumphmodular.com
www.twinmodular.com (trailers)
www.modtech.com (trailers)
www.willscot.com (trailers)
www.pacvan.com (trailers)
www.workspaceplus.com (trailers)
www.usedmodularbuildings.com
www.mbconcepts.com (trailers)
www.triumphmodular.com (trailers)
www.daccotrailers.com (trailers)
www.smartspacemod.com
www.anchormodular.com
www.goldenofficetrailers.com
www.jobsitemobileoffices.com
www.marklineindustries.com
www.uniteam.com.au
www.uniteamchina.com

CONTAINER FRAMES

www.kleusberg.de
www.ofra.de
www.alho.de
www.modbau.de
www.algeco.de
www.strunz.de
www.induo.de (timber frame)
www.cadolto.com
www.eberhardt.eu
www.voestalpine.com
www.smartspacemod.com
www.marmolradzinerprefab.com
www.yorkon.co.uk
www.sekisuiheim.com

Appendix

ORGANIZATIONS

www.ansi.org
ANSI – American National Standards Institute

www.bic-code.org
ICB – International Container Bureau

www.containerhandbuch.de
Cargo Loss Prevention Information from German Marine Insurers

www.din.de
DIN – Deutsches Institut für Normung

www.iata.org
IATA – International Air Transport Association

www.imo.org
IMO – International Maritime Organization

www.imo.org
CSC – The International Convention for Safe Containers

www.marisec.org
ICS – International Chamber of Shipping

www.tis-gdv.de
TIS Transport Informations-Service – Fachinformationen der deutschen Transportversicherer

www.uic.asso.fr
UIC – International Union of Railway

www.unece.org
UN/ECE – United Nations Economic Commission for Europe

REGULATORY WORKS

STANDARDS

ISO 668
Series 1 freight containers, classification external dimensions and ratings
[Amd. 1993 (E)]

ISO 830
Terminology in relation to freight containers, (Amd. 1988)

ISO 1161
Series 1 freight containers – corner fittings Specification (Amd. 1990)

ISO 1496-1
Series 1 freight containers – specification and testing Part 1: general cargo containers for general purposes (Amd.2 – 1998)

ISO 3874
Series 1 freight containers – handling and securing

ISO 6346
Freight containers coding, identification and marking – 1995 (E)

ISO 8323
freight containers – Air/surface (intermodal) general purpose containers – specification and tests

ISO 8501-1
Preparation of steel substrates before application of paints and related products

ISO 12944
Paints and varnishes – corrosion protection of steel structures by protective paint systems

GUIDELINES

Bundesgütegemeinschaft Montagebau und Fertighäuser e.V. (Federal quality association for prefabricated buildings), Germany

RAL quality mark
steel construction system

DIN 18800-7 and 18808
Major welding certificates

Regelmäßige Kontrollen und Optimierung der Fertigung, Qualitätsstandard nach DIN EN ISO 9001 (Regular checking and optimization of production, quality standard as per DIN EN ISO 9001) (TÜV certificate, DEKRA Certification GmbH, Germany)

Surveys on the fulfillment of standards regarding statics, fire protection, noise protection

RGST 1992: Guidelines relating to the process of applying for and receiving permission for the transportation of bulky and heavy loads

EnEV 2009, Energy Savings Ordinance for Buildings, Germany

GLOSSARY

20-FOOT CONTAINER
The most commonly used container size worldwide in the ISO dimensions l/w/h 20'/8'/8'6".

40-FOOT CONTAINER
Basic module for the standardization of ISO containers. Today, it is the most commonly used container size after the 20-foot container.

BRIDGE FITTING
Component for joining containers horizontally.

BSE
Building services equipment; i.e. heating system, electrical installation, sanitation, etc.

BUILDING SITE CONTAINER
Standardized building module with container dimensions based on the model of the freight container, but developed solely for building purposes.

CHASSIS
Undercarriage for supporting transport loads; a term mostly used in vehicle manufacture.

CLIMATE SHELL
Dividing envelope that also acts as an insulator by slowing heat exchange due to its heat transmission resistance.

CONTAINER DIMENSIONS
Typical dimensions of containers according to the ISO standard, refers particularly to length dimensions quoted in units of feet (10 feet, 20 feet, 40 feet, etc.).

CONTAINER FRAME
Further development of existing container types with a consistent separation between frame and filling.

CONTAINER LOOK
Architecture with replicated containers, i.e. conventionally constructed building components, the shape of which is reminiscent of containers.

COR-TEN STEEL
Very strong and weather-resistant steel alloy (CORrosion resistance and TENsile strength).

CORNER FITTING
Attachment points at all corners of a container. They serve as fastening points when lifting, stacking and securing a container.

CROSS MEMBER
Carrier beams made of steel profiles in the base of a container, which support transport loads.

DYNAMIC LOADS
Loads caused by movements (impact forces caused by sliding or loading of containers, for example).

ENEV
German Energy Saving Ordinance; regulations valid in Germany for the reduction of CO^2 emissions from buildings; the ordinance prescribes insulation and heating standards for new buildings.

FILLING
Flat, space-delimiting elements fitted to the supporting structure.

FLEXIBILITY
Ability to adapt to change.

FOOT
Imperial measurement unit originally related to the dimensions of the human foot. The metric conversion factor for the standard English foot is 1 foot = 0.3048 meters. Other derived measurement units are:
1 yard = 3 feet
1 foot = 12 inches
1 inch = 25.4 millimeters

FRAME
Supporting structure of a container, three-dimensional.

FREIGHT CONTAINER
General terms for reusable containers for transporting all types of goods.

GALVANIZATION
Process for providing corrosion protection for steel; the hot-dip galvanization process gives steel a surface coating of zinc; damaged areas must be treated again (cold galvanization).

GLUED-LAMINATED TIMBER
Extremely stable wood building material made of glued boards.

GOOSENECK CHASSIS
Low-based container chassis with a swan-neck coupling component that lowers the loading area height and improves road handling.

GRID
Linear layout pattern for buildings based on the repetition of common distances; there are numerous grid types, which can be identified based on geometry (line grids, band grids, etc.) or classification (structural grids, fittings grids, etc.); the definition of grids/building axes makes unique spatial classification possible.

HIGH CUBE
ISO container with a height of 9 feet and 6 inches; further standardized container heights are the standard cube (8 feet 6 inches) and the square-shaped low cube (8 feet).

HOT GALVANIZING
Corrosion protection by means of thermally applying a zinc coating on steel.

INDIVIDUALIZATION
Adaptation to specific requirements, specialization.

INSULATION
Material with a high heat transmission resistance that is used to reduce heat losses from a building.

ISO
Abbreviation for the International Organization for Standardization, which prepares internationally valid technical "Normung" standards.

ISO CONTAINER
Freight container that fulfils the specifications of the "ISO" ISO 668 standard.

LASHING
Securing systems for fastening containers during transport.

LOW CUBE
ISO container with a reduced height compared to the standard cube (8 feet).

MOBILITY
Movability, not tied to a fixed location.

MODULE
Individual unit in an overall system.

MODULE BUILDING FRAME
Container-like building system using prefabricated cells made of steel frames that can be used as building components; the dimensions deviate from those of containers, however.

NOMINAL DIMENSION
Dimension for orientation that does not correspond with the actual size.

PIN FOUNDATION
Releasable point foundations made from prefabricated concrete parts.

PLYWOOD
Wooden panels consisting of a number of veneer layers bonded to each other in alternating directions at right angles.

PONTOON
Floating concrete or steel body on water that supports a building.

PREFABRICATION
Manufacturing process for building components that takes place in the workshop rather than on site; the greater the degree of prefabrication, the less work remains to be done on site.

PRIMARY SUPPORT STRUCTURE
Main components of the supporting structure that are largely responsible for carrying loads.

MAKESHIFT SOLUTION
Temporarily installed.

OPEN SYSTEM
Building system that is combinable with building components of other systems (e.g. with commercially available building products).

POWERPACK
Compact supply module for building services to achieve technical self-sufficiency in a mobile building.

RAL
Reichs Ausschuß für Lieferbedingungen; Internationally valid quality system with color samples to aid the selection of suitable colors.

RAW CONTAINER
Unconverted version of a container.

REINFORCEMENT
Component that ensures the stability of a body under horizontally acting loads.

REMOUNTABLE
A remountable building can be reassembled again after being disassembled.

RGST
Guidelines valid in Germany relating to the process of applying for and receiving permission for the transportation of bulky and heavy loads.

SECONDARY SUPPORTING STRUCTURE
Secondary components (e.g. secondary carrier beams) that also carry loads—e.g. to provide reinforcement.

SELF-CONTAINED BUILDING SYSTEM
Building system that can only be combined with building components from its own system.

SPATIAL CELL
Building module that already has a load-bearing and surface-defining structure.

STANDARDIZATION
Adapted to generalized requirements, commonization.

STANDARD CUBE
ISO container with a standard height of 8 feet 6 inches.

STATIC LOADS
Loads caused by self-weight (e.g. self-weight of the container, of contents, weight of persons, etc.).

SYSTEM-COMPATIBLE
Complying with the regularities of the modular co-ordinations for containers.

SYSTEM OF UNITS
Uniform measurement system that provides a uniform basis for a certain purpose—e.g. the metric system (definition of the "meter" as a unit of length) or the system based on the foot.

TEMPORARY
Provisional, for a short time.

TEU
Abbreviation for Twenty-foot Equivalent Unit, an internationally standardized unit for describing the manufacturing, loading and moving capacities of freight containers; derived from the size of a 20-foot ISO container.

THERMAL SEPARATION
Reduction of the thermal flux by means of a lower heat-conducting material within a building component.

THERMAL BRIDGES
Weak points in a structure that cause increased heat losses.

TOLERANCE DIMENSION
Dimension defines the space between two assembled containers; it serves to compensate building tolerances.

TRAILER
A building site container with a chassis widely used in the USA.

TRANSPORTABLE
Transportable units have devices for transport, but do not have a chassis or similar.

TRAPEZOIDAL SHEET
Corrugated sheet manufactured from cold-rolled steel with a profile depth sufficient to provide load-bearing capacity; used as wall paneling on steel containers.

TWISTLOCK
Joining component for vertically connecting containers to each other and to the base surface in a secure, interlocking manner.

TYPE STATICS
Typical statics calculation for the load-bearing capacity of a container that is only valid for this particular model in this form; new statics calculations are necessary if modifications are carried out.

VAPOR BARRIER
Sheeting that reduces the transmission of water molecules in the form of vapor; necessary in construction to prevent the diffusion of saturated warm air from the building interior into the components of the climate shell, as otherwise vapor would condense and cause dampness in these components.

VARIABILITY
Variety of solutions in different variants.

Appendix

INDEX OF PROJECTS

ADAM KALKIN
USA, www.architectureandhygiene.com

12 CONTAINER HOUSE // PAGE 148
Engineer's name: Butler Corp
Client: Anne Adriance, Matt Adriance
Completion date: 2002
Total floor area: 372 m²
Number of containers: 12
Construction costs: € 358,000
Photographer: Peter Aaron/Esto Photographics Inc.

ILLY CAFÉ // PAGE 56
Engineer's name: Quik Build
Client: Illy Coffee
Completion date: 2007
Number of containers: 1
Photographer: Luca Campigotto

OLD LADY HOUSE // PAGE 145
Engineer's name: Chrystle Engineering
Client: Various
Completion date: 2008
Total floor area: 93 m²
Number of containers: 3
Construction and materials: steel, glass
Photographer: Peter Aaron/Esto Photographics Inc.

PUSH BUTTON HOUSE // PAGE 57
Engineer's name: Quik Build
Client: ILLY Coffee
Completion date: 2005
Number of containers: 1
Photographer: Peter Aaron/Esto Photographics Inc.

SUBURBAN HOUSE KIT // PAGE 144
Engineer's name: Anderson engineer
Client: Deitch Projects
Completion date: 2004
Number of containers: 5
Photographer: Peter Aaron/Esto Photographics Inc.

AFF ARCHITEKTEN
Germany, www.aff-architekten.com

SANITARY FACILITIES FOR SUMMER CAMP // PAGE 94
Client: Hochbauamt Magdeburg
Completion date: 2002
Total floor area: 82 m²
Building Volume: 208 m³
Number of containers: 4
Structural system: container box tubing/filled-in frame construction, exterior trapezoidal sheet paneling
Construction and materials: 120 mm mineral wool insulation, inner planking, melamine resin-coated MDF panels (walls, ceiling), textured surface, floor: cement fiber plate, epoxy resin, coated
Construction costs: € 115,000
Photographer: Archiv AFF Architekten, Sven Fröhlich

ALLFORD HALL MONAGHAN MORRIS
Great Britain, www.ahmm.co.uk

RAINES COURT // PAGE 174
Engineer's name: Whitby Bird
Client: Peabody Trust
Completion date: 2003
Construction costs: GBP 8.9 million

ANGELA FRITSCH ARCHITEKTEN
Germany, www.af-architekten.de

GOLD PAVILION // PAGE 95
Client: Alice Hospital, Darmstadt
Completion date: 2005
Total floor area: 48 m²
Building Volume: 181 m³
Number of containers: 1
Structural system: steel-frame construction
Construction and materials: exterior: 3 mm lasered sheet aluminum, interior: gypsum board, wood laminate
Construction costs: € 68,000
Photographer: Dieter Leistner

KING KAMEHAMEHA BEACH CLUB // PAGE 152
Client: Kingkameha Club GmbH
Completion date: 2004
Total floor area: outdoor facilities: 10,200 m²
Containers: 13.4 m² each
Building Volume: 31.6 m³ ea. Container
Number of containers: 27
Structural system: steel wall system, used freight containers
Construction and materials: exterior: painted, interior: fitted with multi-layer wood panels
Construction costs: € 200,000
Photographer: Dieter Leistner

ARTUR LESCHER, BERNARDES+JACOBSEN
Italy, www.containerart.org

CONTAINERART SÃO PAULO // PAGE 136
Client: Automatica
ContainerArt partner: Daniel Roesler Global project
Coordination: Ronald Lewis Facchinetti
Completion date: 2008
Total floor area: 825 m²
Photographer: Paulo Rodrigo Sousa Grangeiro Local

ATELIERWORKSHOP
New Zealand, www.atelierworkshop.com

PORT-A-BACH // PAGE 50
Engineer's name: Spencer-Holmes
Client: Sam Saffery
Completion date: 2007
Total floor area: 37 m²
Building Volume: 100 m³
Number of containers: 1
Structural system: container steel, stainless steel
Construction and materials: bamboo plywood, wool insulation, canvas/pine, decking/aluminium doors and windows
Photographer: Paul McCredie

BIJVOET ARCHITECTUUR & STADSONTWERP
the Netherlands, www.wijnofwater.nl

WIJN OF WATER // PAGE 108
Engineer's name: Ir. Caroline Bijvoet
Client: Bureau Medelanders
Completion date: 2005
Total floor area: 217 m² (exclusive of terrace)
Building Volume: 629 m³
Number of containers: 9
Structural system: 40-foot containers, advice about maximum size of holes without loss of containers stability/strength: Leen Brak in Gouda
Construction and materials: built by Ap Klein, small contractor; construction only of the containers themselves, insolated with gypsum board only (construction)
Construction costs: € 165,000
Building services engineering: DH installaties
Photographer: Maarten Laupman

BUREAU DES MÉSARCHITECTURES
DIDIER FIUZA FAUSTINO
France, www.mesarchitecture.com

SKY IS THE LIMIT // PAGE 76
Completion date: 2008
Total floor area: 100 m²
Construction and materials: steel, varnished plywood, glass, steel duckboards
Photographer: Hong Lee (05); all other photographs: Bureau des Mésarchitectures

CASTOR DESIGN
BRIAN RICHER AND KEI NG
Canada, www.castordesign.ca

SAUNA BOX // PAGE 100
Client: many
Completion date: 2005
Total floor area: 6 m²
Building Volume: 14.5 m³
Number of containers: 1
Construction and materials: cedar, metal, glass, marble, bronze, solar panels, wood fired stove
Construction costs: $ 25,000 (retail price)
Photographer: castor

DEMARIA DESIGN ASSOCIATES INC.
PETER DEMARIA WITH CHRISTIAN KIENAPFEL
USA, www.demariadesign.com

PARSONAGE BOYLE HEIGHTS // PAGE 227
Architect: Peter DeMaria, AIA Architekt, Dipl.-Ing. Christian Kienapfel, Architekt, LEED AP

Appendix

Client: Church Municipality
Completion date: 2009
Total floor area: 390 m²
Number of containers: 7
Structural system: cargo container and concrete/steel construction
Building services engineering: HVAC
Photographer: Andre Movsesyan

REDONDO BEACH HOUSE // PAGE 228
Architect: Peter DeMaria, AIA Architekt, Dipl.-.Ing. Christian Kienapfel, Architekt, LEED AP
Client: private client
Completion date: 2008
Total floor area: 270 m²
Number of containers: 8
Structural system: cargo container and conventional stick frame constuction (hybrid)
Construction and materials: airplane hangar doors, formaldehyde free plywood, recycled denim insulation/cotton fiber insulation, multi-skinned acrylic sheets, ceramic insulation
Construction costs: $ 427,000
Building services engineering: FAU
Photographer: Andre Movsesyan

EER ARCHITECTURAL DESIGN
GEERT BUELENS AND VEERLE VANDERLINDEN
Belgium, www.eerdesign.com

CONTAINING LIGHT MOBILE UNIT // PAGE 60
Engineer's name: Dwen; Ney & Partners
Client: Kreon
Completion date: 2005
Total floor area: 30 m²
Building Volume: 80 m³
Number of containers: 1
Structural system: 4 hydraulic cylinders
Construction and materials: steel
Construction costs: € 200,000
Building services engineering: control by AIB vinçotte
Photographer: Serge Brison

FPS OFICINA DE ARQUITECTURA
FRANCISCO FENILI, JORGE PÉREZ GONZÁLEZ AND JULIO SEPIURKA
Argentinia, www.fpsarquitectura.com

X-SPACE ATELIER // PAGE 80
Engineer's name: Enrique Ibañez Arq.
Client: Rosa Skific
Completion date: 2007
Total floor area: interior 18 m², exterior 8 m²
Building Volume: 54 m³
Number of containers: 1
Structural system: steel frame
Construction and materials: steel frame, steel skin and gypsum board inside, windows: aluminium, floor: wood
Building services engineering: Carlos Guayapero
Construction costs: US$ 10,000
Photographer: FPS Oficina de Architectura S.H.

FRADE ARQUITECTOS
JUAN PABLO RODRÍGUEZ FRADE
Spain, www.fradearquitectos.com

BOOK FAIR PAVILION // PAGE 153
Client: Ayuntamiento de Madrid
Completion date: 2004
Total floor area: 173 m²
Building Volume: 850 m³
Number of containers: 11

Construction costs: € 100,000
Building services engineering: air condicionated
Photographer: LA NAVE.

GARY DEIRMENDJIAN
Australia, www.garo.com.au

CONSUMER TEMPLE – BROKEN ICON // PAGE 135
Client: invited artist, Armory Exhibition 09
Completion date: 2009
Total floor area: 12.25 m²
Building Volume: 98 m³
Photographer: Gary Deirmendjian

GUSTAU GILI GALFETTI
Spain, www.gustaugili.com

CONTAINER HOUSING // PAGE 64
Client: APTM- Construmat_ Fira de Barcelona
Completion date: 2005
Total floor area: 30 m²
Number of containers: 1
Photographer: Oriol Rigat

HAN SLAWIK
Germany, www.slawik.net

BED BY NIGHT // PAGE 162
Engineer's name: Alexander Furche
Client: State capital city of Hanover / Office of Children's and Family Affairs
Completion date: 2002
Total floor area: 361 m²
Building Volume: 1,542 m³
Number of containers: 19
Structural system: wooden supporting structure of laminated timber
Construction and materials: premounted industrial glass facade
Construction costs: € 192,000
Photographers: Karl Johaentges (03, 08), Han Slawik (01, 02, 06, 09)

CAMPUS // PAGE 184
Client: Studio voor Architectuur en Techniek, Almere, the Netherlands
Completion date: 1992
Number of containers: 5
Structural system: five vertically positioned 20-foot shipping containers, combined with industrially prefabricated building products
Photographer: Han Slawik

FLOATING HOUSE // PAGE 86
Completion date: 2008 (draft)
Total floor area: 92 m²
Building Volume: 429 m³
Number of containers: 10 moduls

HOMEBOX1 // PAGE 92
Engineer's name: overall stability: Prof. A. Furche, Köngen, Germany, Furche Zimmermann structural engineers, Containers: Prof. E. Papsch
Completion date: 2009
Total floor area: 14.35 m²
Building Volume: 42 m³
Number of containers: 1
Structural system: wooden containers with glued laminated timber elements with ISO corners
Construction and materials: insulation and multi-layer panels as facade structure, wooden windows
Building services engineering: self-sufficient supply

and disposal: tanks for freshwater, rainwater and wastewater, photovoltaic elements
Construction costs: € 25,000
Photographer: Frank Aussieker/Architekturfotografie

MYSTERY CUBE // PAGE 186
Client: Niedersächsisches Landesmuseum Hannover
Completion date: 2002
Total floor area: 52 m²
Building Volume: 295 m³
Number of containers: 4
Structural system: stacked building containers
Construction and materials: standard construction scaffolding surrounded by integrated stairs, covered with a black fiber-glass fabric
Construction costs: € 15,700 + € 390 for equipment

HAN SLAWIK
WITH IMS INGENIEURGESELLSCHAFT MBH AND BOF ARCHITEKTEN

IBA DOCK // PAGE 214
Realisation: bof architekten, Bücking Ostrop Flemming
Engineer's name: IMS Ingenieurgesellschaft mbH
Client: IBA Hamburg GmbH, Internationale Bauausstellung
Completion date: 2010
Total floor area: 1,831 m²
Building Volume: 6,855 m³
Number of containers: 36 x 60 foot containerframes
Structural system: modular construction with steel frames
Construction costs: € 8 million
Photographer: Rüdiger Mosler, Mainz, KLEUSBERG Modul- und Systembau, www.kleusberg.de (05, 06);
Rendering: BloomImages (03, 04)

HELEN & HARD ARCHITEKTEN
Norway, www.hha.no

B-CAMP // PAGE 156
Engineer's name: Dimensjon Rådgivning
Client: Helen & Hard ANS
Completion date: 2004
Total floor area: 140 m²
Building Volume: 380 m³
Number of containers: 8
Structural system: structural containers stacked on top of each other
Construction and materials: dwelling containers from offshore industry, fitted with second hand windows and doors. cladding: transparent corrugated plastic panels, and scrap metal from metal industry.
Construction costs: € 200,000
Photographer: Emile Ashley

HSH ARCHITECTS
Czech Republic, www.hsharchitekti.cz

RESIDENTIAL CONTAINERS // PAGE 78
Client: private person
Completion date: 2009
Total floor area: 151.6 m²
Structural system: wooden frame system cladded with titan zinc sheets
Construction and materials: local materials, wooden boards, metal panels, plasterboard, glass
Construction costs: € 120,000
Photographer: Ester Havlová

Appendix

HVDN ARCHITECTEN
the Netherlands, www.hvdn.nl

QUBIC AMSTERDAM // PAGE 176
Engineer's name: HVDN architecten
Client: De key, de principaat, amsterdam
Completion date: 2005
Total floor area: 31,858 m²
Building Volume: 94,021 m³
Number of containers: ≤ 1,040
Structural system: steel
Construction and materials: steel, synthetic material
Building services engineering: advised by Traject Advies groep, www.traject.com
Construction costs: € 18.5 million
Photographer: Luuk Kramer

HYBRID, CARGOTECTURE
USA, www.hybridseattle.com

C320S // PAGE 96
Engineer's name: Sliderole Engineering
Client: Alexander Farms
Completion date: 2004
Total floor area: 30 m²
Building Volume: 86 m³
Number of containers: 2
Structural system: modules as shear, envelope, gravity loads, and entirety of enclosure.
Construction and materials: steel containers, tapered foam roof insulation, batt insulation, relocatable beam-and-footing foundation, wood deck, plywood interiors.
Construction costs: $ 50,000
Building services engineering: solar panels, roof water collection, green machine (relocatable septic system), heat recovery ventilator, dimmable lighting, toilet flushes collected water.
Photographer: Lara Swimmer
Drawings: HyBrid

IC GREEN INC.
USA, www.icgreen.net

BRENTWOOD CABANA // PAGE 68
Engineer's name: Parker Resnick
Client: M. Kelly
Completion date: 2009
Total floor area: 30 m²
Building Volume: 71 m³
Number of containers: 2
Structural system: modified container structure on pier foundations
Construction and materials: fritted glass partitions, glass tiles, wood benches and cabinets, fabric curtains, stainless steel mesh curtain
Building services engineering: electric and plumbing only
Photographer: IC Green

IMS INGENIEURGESELLSCHAFT MBH AND BOF ARCHITEKTEN
Germany, www.bof-architekten.de, www.ims-ing.de

RESEARCH STATION // PAGE 220
Engineer's name: m+p consulting, Braunschweig
Client: NCAOR/National Centre for Antarctic & Ocean Research, Goa, India
Completion date: 2011/2012
Total floor area: 2,150 m²
Building Volume: Including outer shell: 7,850 m³
Number of containers: 132 containers (20 feet), the corridors in the upper stories consist of "flats"
Structural system: containers are used to provide the supporting structure. The protruding part of the station has an additional steel supporting structure with V supports.
Building services engineering: the building has three combined heat and power units for back-up purposes that generate the electricity required. The waste heat from the CHP units is harnessed and used to heat the station. In addition, the station is fitted with ventilation equipment, and the air-conditioning maintains an air humidity of around 30%. Drinking water is produced from seawater using a desalination unit, and waste water treatment is also carried out.

JAMES & MAU ARCHITECTS
Chile, www.jamesandmau.com

INFINISKI MANIFESTO HOUSE // PAGE 104
Client: private
Completion date: 2009
Total floor area: 160 m²
Construction costs: € 550/m² in Chile. € 750/m² in Europe
Photographer: Antonio Corcuera

JAN HOHLFELD, CLAUS HESEMANN, MARKO HEINSDORFF
HERIBERT WEEGEN, UWE HARZER AND ANNE MEYER
Germany, www.koelnerbox.de

KOELNERBOX (PROTOTYPE) // PAGE 88
Architects Design: Jan Hohlfeld, Claus Hesemann, Marko Heinsdorff
Architects realization: Heribert Weegen, Uwe Harzer, Anne Meyer
Client: JACK IN THE BOX e.V.
Completion date: 2007
Total floor area: 30 m²
Building Volume: 90 m³
Number of containers: 1
Structural system: container support system not modified; reinforcements around the additional openings
Construction and materials: windows/doors: steel frame/acrylic glass, insulation: mineral wool, inner paneling: glazed or coated blockboard
Building services engineering: optional electrical/data connection in each extension module, natural ventilation
Construction costs: € 25,000
Photographer: JACK IN THE BOX/Thomas E. Albrecht

JEAN NOUVEL
WITH ZIEBELL & PARTNER
France, www.jeannouvel.com

WISMAR TECHNOLOGY AND RESEARCH CENTER // PAGE 180
Ingenieure: INROS Planungsgesellschaft Rostock (supporting structure), Ingenieurbüro F. Barkowski (water), Ingenieurbüro R. Goosmann (electricity)
Client: technology and business center, Mr Klaus Seehase
Completion date: 2003
Total floor area: 5,500 m²
Construction costs: € 13 million
Photographer: Philippe Ruault

KEITH DEWEY
Canada, www.zigloo.ca

ZIGLOO DOMESTIQUE // PAGE 226
Engineer's name: Ritchie Smith (Hoel Engineering)
Client: the Dewey family
Completion date: 2006
Total floor area: 178 m²
Building Volume: 435 m³
Number of containers: 8 (20 feet long) + 1 (10 feet long)
Structural system: container's structure
Construction and materials: steel, glass, chain-link fence, bamboo, concrete, spray-foam insulation, wood, quartz-composite countertop, drywall
Building services engineering: structural engineering: Hoel Engineering, building envelope specialist: Chatwin Engineering, builder: David Gauld
Construction costs: CDN$ 350,000
Photographer: Nik West

KNOCK
Sweden, www.knock.se

VOLVO C30 EXPERIENCE PAVILION // PAGE 198
Engineer's name: Space Display
Client: Volvo Cars, www.volvocars.com
Completion date: 2006
Total floor area: 468 m²
Number of containers: 54 and a 50 m long hallway where the guests entered the event
Photographer: KNOCK, Mikael Olsson

KOMMA4 ARCHITEKTEN, PÜTZ/REETZ
Germany, www.komma4.net

TEMPORARY EXHIBITION / 100 YEARS OF FOOTBALL CLUB ST. PAULI // PAGE 114
Engineer's name: Peper GmbH
Client: FC St. Pauli soccer club, Hamburg
Completion date: in progress
Total floor area: 800 m²
Building Volume: 3,700 m³
Number of containers: 44
Structural system: frame structure of the containers
Construction and materials: various fittings for the presentation of exhibition pieces
Building services engineering: ventilation fans
Construction costs: € 250,000

LEIBNIZ UNIVERSITY HANOVER
HILDE LÉON AND UDO WEILACHER
Germany, www.entwerfen.uni-hannover.de/leon/

HALF MILE // PAGE 127
Architects: Leibniz University Hanover, Faculty of Architecture and Landscape Sciences, Prof. Hilde Léon and Prof. Udo Weilacher
Completion date: 2006
Total floor area: 15 m²
Building Volume: 90 m³
Number of containers: 2

LHVH ARCHITEKTEN
Germany, www.lhvh.de

DERCUBE // PAGE 187
Client: derCube AG/KulturFlecken Silberstern e.V., Freudenberg
Completion date: 2007

Total floor area: 400 m²
Building Volume: 1,500 m³
Number of containers: 36
Structural system: steel frame structure (30 containers, 6 stair towers, ALHO Systembau-Basic line)
Construction and materials: standard construction dry mortarless construction, HPL boardan walls and ceelings, PVC floors, EPS pressure cloth front
Building services engineering: Standard power and water supply, structural sunscreen by pressure cloth front
Construction costs: €495,000
Photographer: Axel Ollenschläger

INTERSTITIAL LIVING // PAGE 115
Client: F. Lohner, J. Voss, F. Holschbach
Completion date: 2005
Total floor area: 90 m²
Building Volume: 180 m³
Number of containers: 2
Structural system: steel-frame supporting structure (2 basic line construction site office containers, 1 basic line floor frame, ALHO system structure)
Construction and materials: Facade: standard profile sheets, mullion-and-transom glazing, smoothed and painted drywall on walls and ceilings, painted oriented strand boards for flooring
Building services engineering: standard electricity and water supply
Construction costs: €40,000
Photographer: ALHO Systembau/Mr Schade

MEETINGPOINT PLAN06 // PAGE 126
Client: plan project, Sabine Voggenreiter and Kay von Keitz GbR
Completion date: 2006
Total floor area: 100 m²
Building Volume: 255 m³
Number of containers: 9
Structural system: steel-frame supporting structure (6 construction site office containers, 3 stair towers, ALHO Systembau-Basic line)
Construction and materials: sandwich panels and full-length windows as facade, drywall construction, high-pressure laminate sheets on the walls and ceilings, PVC floor covering
Building services engineering: standard electricity supply
Construction costs: €60,000
Photographer: ALHO Systembau/Mr Schade

SEVEN-DAY ARCHITECTURE // PAGE 188
Client: ALHO Systembau, Friesenhagen
Completion date: 2004
Total floor area: 725 m²
Building Volume: 2,610 m³
Number of containers: 48
Structural system: steel-frame supporting structure (29 construction site office containers, 17 frames, 3 stairwell towers, ALHO Systembau-Basic line)
Construction and materials: 1st floor and 2nd floor: Standard drywall construction, high-pressure laminate sheets on the walls and ceilings, PVC floor; 3rd floor: finishing in accordance with the German energy saving ordinance standard, drywall construction with corresponding insulation on the ceilings and walls, linoleum floor, full-length window, cloth facade
Building services engineering: standard electricity and water supply, in-built sun protection from cloth facade
Construction costs: €495,000
Photographer: ALHO Systembau/Mr Schade

LOT-EK
USA, www.lot-ek.com

BILLBOARD BUILDING // PAGE 217
Engineer's name: Robert Silman Associates
Client: Goldman Properties
Completion date: in progress
Total floor area: 1,022 m²
Number of containers: 30
Structural system: shipping containers, steel + glass base
Renderings: LOT-EK

BOHEN FOUNDATION // PAGE 54
Engineer's name: Ove Arup
Client: Fred Henry/Bohen Foundation
Completion date: 2002
Total floor area: 1,394 m²
Number of containers: 8
Construction costs: $1.5 million
Photographers: Nicholas Koenig (01, 02, 03, 06), Paul Warchol Photography Inc. (04, 05, 07)

MDU / MOBILE DWELLING UNIT // PAGE 53
Engineer's name: engineering + fabrication: UAF Construction/Marc Ganzglass
Completion date: 2003
Total floor area: 46 m²
Structural system: modified shipping container
Construction and materials: shipping container, plywood, nudo boards, fluorescent lights
Construction costs: $150,000
Number of containers: 1
Photographer: courtesy of the Walker Art Center

PIER 57 // PAGE 216
Engineer's name: structure: Robert Silman Associates
MEP + sustainability: Buro Happold
Client: YoungWoo & Associates
Completion date: in progress
Total floor area: 3,160 m²
Number of containers: approximately 500
Structural system: refurbishment and reuse of existing steel and concrete Pier structure + addition of shipping containers architecture
Renderings: LOT-EK

PUMA CITY // PAGE 140
Engineer's name: structure: Robert Silman Associates, MEP: David Rosini
Client: PUMA
Completion date: 2008
Total floor area: 1,020 m²
Number of containers: 24
Structural system: reinforced shipping containers
Construction and materials: shipping containers, apitong wood, galvanized steel stairs, LED lights
Photographer: Danny Bright

SANLITUN SOUTH // PAGE 218
Engineer's name: Beijing Architectural & Engineering Design Company
Client: Guo Feng
Completion date: 2008
Total floor area: 24,000 m²
Number of containers: 151
Structural system: reinforced concrete structure with modified shipping containers
Photographer: Shuhe Architectural Photography

UNIQLO FLAGSHIP // PAGE 66
Engineer's name: Shimizu Building Life Care Kansai
Client: UNIQLO
Completion date: 2004
Total floor area: 1,022 m²
Number of containers: 1

UNIQLO POP-UP STORES // PAGE 66
Engineer's name: engineering + fabrication: TRS Containers
Client: UNIQLO
Completion date: 2006
Total floor area: 30 m²
Number of containers: 1 per unit
Structural system: modified shipping container
Construction and materials: shipping container, plywood, nudo board shelves, fluorescent lights, retractable duct dressing room
Construction costs: €29,000
Photographer: Danny Bright

LUC DELEU
Belgium, www.topoffice.to

CONSTRUCTION X // PAGE 134
Engineer's name: Dirk Jaspaert
Client: Middelheim open air sculpture museum, Antwerp, Belgium
Completion date: 2003
Number of containers: 9
Construction and materials: containers, twistlocks, screw bridge fittings, stacking cones, special interconnecting devices
Photographer: Philippe de Gobert

HOORN BRIDGE // PAGE 134
Engineer's name: Dirk Jaspaert
Client: organization of the exhibition "For real now"
Completion date: 1990
Total floor area: 30 m²
Number of containers: 2
Construction and materials: containers (back wall removed), screw bridge fittings, special interconnecting devices
Photographer: Wim Riemens

ORBINO // PAGE 132
Engineer's name: Dirk Jaspaert
Completion date: 2004
Total floor area: 45 m²
Construction and materials: containers (three remodeled containers), screw bridge fittings, stacking cones, steel staircase, concrete base
Photographer: Han Slawik

SPEYBANK // PAGE 134
Engineer's name: Dirk Jaspaert
Client: open air sculpture, Museum Middelheim, Antwerp, Belgium
Completion date: 2003
Number of containers: 4 x 20-foot containers
Construction and materials: containers, screw bridge fittings, stacking cones, concrete base, special interconnecting devices.
Photographer: Philippe de Gobert

LUIS DE GARRIDO
Spain, www.luisdegarrido.com

R4HOUSE // PAGE 164
Client: Particular. Exposed at International Faer Construmat 2007, Barcelona
Completion date: 2007 (made in only 20 days)
Total floor area: large model: 150 m², small model: 30 m²
Number of containers: 6
Construction costs: large model: €60,000, small model: €12,000

Appendix

LUIS RODRÍGUEZ ALONSO, JAVIER PRESA
Spain, www.habitainer.com

KIOWA PROTOTYPE (HABITAINER) // PAGE 52
Completion date: 2006
Total floor area: 14 m²
Building Volume: 30 m³
Number of containers: 1
Structural system: container structural chasis, self-standing dinamic groundation feet
Construction and materials: steel 20-foot container 1BB. Internal panel system composite on polyurethane, polystyrene, wood frame and Medium Density recycled fiber board. White matte paint finish on ceiling and walls. Plastic continuous floor finish. Electricity protection board + 3 lines interior conductions. WC, toilet and shower units. Steel plumbing exterior conduction.
Construction costs: € 8,000
Photographer: Pablo Rodríguez

MARMOL RADZINER PREFAB
USA, www.marmol-radziner.com, www.marmolradzinerfurniture.com

HIDDEN VALLEY // PAGE 234
Client: private
Completion date: 2007
Total floor area: indoor 230 m², terraces 160 m²
Number of containers: 15 moduls
Structural system: recycled-steel frame modules, made in a factory, shipped complete
Construction and materials: main house: 12 recycled steel-framed modules, garage: 3 modules, metal panel siding, ipe wood decks, sadlestone floors, walnut casework, caesarstone countertops, geothermal ground loop HVAC and pool heating system, kw solar planel array backup generator, 900 feet well and 1700 gallon cistern
Building services engineering: general contractor: Marmol Radziner, mechanical engineer: Trey Austin, electrical engineer: Nikolakopulos, structural engineer: C.W.Howe
Photographer: Joe Fletcher

MEYER EN VAN SCHOOTEN ARCHITECTEN
the Netherlands, www.meyer-vanschooten.nl

TEMPORARY ACCOMMODATION FOR 1001 DAYS // PAGE 190
Client: Meyer en Van Schooten Architecten, Amsterdam
Completion date: 2000
Total floor area: 800 m²
Building Volume: 5,000 m³
Number of containers: 10 containers + 1 as entrance
Structural system: the formar factory hall was used as the outer skin / space for the project. The structure of the 10 sea containers was used to construct the spaces within the space.
Contractor, builder: Bouwbedrijf E.A. van den Hengel bv, Soest
Construction costs: € 154,800 +/- € 195 per m²
Photographers: Luuk Kramer (03, 04, 05);© Hans Fonk (02, 06)
Drawings: Meyer en van Schooten Architecten

MM+P / MEYER-MIETHKE & PARTNER
Germany, www.modularesbauen.com

50,000 € MODULAR-HOUSE // PAGE 189
Completion date: 2000-2002

MMW ARCHITECTS OF NORWAY
Norway, www.mmw.no

GAD // PAGE 138
Engineer's name: Uniteam AS
Client: Alexandra Dyvi
Completion date: 2006
Total floor area: 258 m²
Building Volume: 774 m³
Number of containers: 10
Structural system: prefabricated freight containers
Construction and materials: steel, glass, tree (plywood), plaster, spray-paint (surface treatment)
Building services engineering: Uniteam AS
Photographer: Erik Førde (02, 03, 05)
Rendering: Jon Arne Jørgensen (01)
Drawings: Sindre Østereng (04, 06, 07, 08)

MVRDV
the Netherlands, www.mvrdv.nl

CANCER CENTER AMSTERDAM // PAGE 178
Engineer's name: De Meeuw
Client: Cancer Center Amsterdam, the Netherlands
Completion date: 2005
Total floor area: 6,000 m²
Number of containers: 256
Photographer: Rob 't Hart

NEST ARCHITECTS
Australia, www.nestarchitects.com.au

GORMAN SHIP SHOP // PAGE 91
Engineer's name: K H Engineering
Client: Gorman Industries
Completion date: 2008
Total floor area: 14.5 m²
Building Volume: 236 m³
Number of containers: 1
Structural system: there is no structural system other than the shipping container.
Construction and materials: the rawness of the container is a key feature of the design, minimal intervention evokes its previous life. All the fit out materials are either recycled or from sustainable sources and designed to sheet sizes to minimise waste. The shop makes use of daylight through porthole windows that also encourage passers-by to peek inside. Compact fluorescent globes and a laptop terminal for online purchases run on green power. The building materials used to fit out the internal space are recycled or sustainable affording the project minimal carbon emissions.
Building services engineering: The steel prefabricated container was chosen due to its ease of transport and because it is basically a carbon neutral object. Hoop pine plywood was used throughout the interior because it is produced from sustainably managed sources. The plywood used uses non-formaldehyde based glues. Moreover, plywood has an excellent weight to strength ratio, in some cases being stronger than steel but lighter, therefore reducing the overall weight of the object and saving in transport borne CO_2 emissions.
Construction costs: AU$ 25,000
Photographer: Jesse Marlow

NICHOLAS LACEY AND PARTNERS
Great Britain, www.containercity.com

TRINITY BUOY WHARF (CONTAINER CITY I) // PAGE 168
Engineer's name: Buro Happold
Client: Urban Space Management Ltd.
Completion date: 2001
Total floor area: 450 m²
Number of containers: 20
Photographer: Teresa Lundquist

TRINITY BUOY WHARF (CONTAINER CITY II) // PAGE 168
Engineer's name: Buro Happold
Client: Urban Space Management Ltd.
Completion date: 2002
Number of containers: 30
Photographer: Teresa Lundquist

O ANTIDOTO
ANTONIA MAIO AND JAVIER QUINTEIRO
Spain, www.behance.net

DINAHOSTING OFFICES // PAGE 62
Client: www.dinahosting.co
Completion date: 2008
Total floor area: 800 m²
Construction and materials: corrugated steel
Construction costs: € 300,000
Photographer: Héctor Fernández Santos-Díez / fabpics

OMD – OFFICE OF MOBILE DESIGN
JENNIFER SIEGAL WITH KELLY BAIR
USA, www.designmobile.com

SEATRAIN HOUSE // PAGE 222
Architects: Jennifer Siegal (Principal), Kelly Bair (Assistant)
Client: Richard Carlson
Completion date: 2003
Total floor area: 279 m²
Number of containers: 2 aluminum shipping containers, 2 steel shipping containers, 2 aluminum grain trailers (pool and pond)
Construction and materials: B-36 steel roof decking, tapered steel beams, recycled-wood joists, aluminum-frame windows, salvage-steel framing, recycled carpet, cherrywood flooring, tube-steel pergola, flagstone water wall
Photographer: Daniel Hennessey

PATZNER ARCHITEKTEN
Germany, www.patzner-architekten.de

BORNACK DROP STOP TRAINING CENTER // PAGE 70
Engineer's name: Wulle Ingenieure
Client: Bornack GmbH+Co. KG
Completion date: 2006
Total floor area: entire building approx. 6,500 m²
Containers approx. 150 m²
Building Volume: Entire building 110,000 m³, containers 375 m³
Number of containers: 10
Construction costs: ca. € 300,000 (conversion costs)
Photographer: Stefan Hohloch

PAUL STANKEY, SARAH NORDBY
USA, www.hivemodular.com

HOLYOKE CABIN // PAGE 98
Engineer's name: Scott Stankey, Krista Stankey
Client: Paul Stankey, Scott Stankey, Krista Stankey, Sarah Nordby
Completion date: 2000
Number of containers: 2
Structural system: steel and wood
Construction and materials: concrete, steel, wood, glass
Construction costs: $ 20,000

Appendix

PHOOEY ARCHITECTS
Australia, www.phooey.com.au

CHILDREN'S ACTIVITY CENTRE // PAGE 110
Engineer's name: Perrett Simpson Consulting Engineers
Client: City of Port Phillip
Completion date: 2007
Total floor area: 127 m²
Number of containers: 4
Structural system: re-used freight container
Construction and materials: material durable, recycled, reclaimed, plantation, reused or salvaged from demolition; this included decking, windows, carpet tiles and joinery. The whole container was re-used including doors suspending balconies. As a process of feedback design, the container off-cuts (with cannibalised super scale graphics) were recycled into the final building. Container off-cuts were re-fashioned into functional balustrades, sun shading and decoration. Timber deck off-cuts became soffit linings.
Construction costs: € 83,000
Photographer: Peter Bennetts

PIERRE MORENCY ARCHITECTE
Canada, pierremorencyarchitecte.com

CHALET DU CHEMIN BROCHU // PAGE 232
Client: Pierre Morency
Completion date: 2007
Total floor area: 160 m²
Number of containers: 3
Structural system: foundation and piloti
Construction and materials: concrete foundation, container, hemlock as a second skin, steel columns to hold the third container
Construction costs: € 85,000
Photographer: Normand Rajotte

PLATOON.BERLIN & SOEREN ROEHRS
Germany, www.platoon.org

PLATOON CULTURAL DEVELOPMENT BERLIN // PAGE 116
Client: PLATOON
Completion date: first status 2002, last 2007
Total floor area: 110 m²
Building Volume: 280 m³
Number of containers: 4 + 1 pool
Structural system: no additional enforcement
Construction and materials: wooden frame work, plywood, filled with cellulose, planked with aluminum cover
Construction costs: € 80,000
Building services engineering: aircondition heating system
Photographer: Christian v. Steffelin

PLATOON AND GRAFT ARCHITECTS
WITH U-IL ARCHITECTS
Germany, www.platoon.org

PLATOON KUNSTHALLE SEOUL // PAGE 192
Architect: PLATOON in collaboration with GRAFT Architects, Beijing + U-il Architects, Seoul, South Korea
Engineer's name: Midas IT, Seoul, South Korea
Client: PLATOON cultural development
Completion date: 2009
Total floor area: 950 m²
Building Volume: 3,600 m³
Number of containers: 28
Structural system: reinforced cargo containers stacked and welded
Construction and materials: frame work, plasterboard, mineral wool, covered with aluminium plates
Building services engineering: main hall floor heating, cooling ventilation by open garage doors; container rooms with cooling & warming air condition; complete sanitary installations + restaurant kitchen
Construction costs: € 715,000
Photographer: PLATOON

PLOT = JDS + BIG
Denmark, www.jdsarchitects.com

SJAKKET YOUTH CENTER // PAGE 81
Engineer's name: Birch & Krogboe
Completion date: 2007
Total floor area: 2,000 m²
Number of containers: 1
Construction costs: € 3 million
Photographers: Vegar Moen (01, 07), Felix Luong (02, 03, 04, 05, 06, 10)

PLUS+ BAUPLANUNG
WITH ENGELHARDT.EGGLER.ARCHITEKTEN.
Germany, www.plus-bauplanung.de

THEATERHAUS STUTTGART // PAGE 74
Engineer's name: supporting structure: Löffler Ingenieur Consult GmbH, existing supporting structure: Dr.Ing. Adrian Pocanschi
Client: Stiftung Pragsattel Theaterhaus
Completion date: 2003
Total floor area: 12,200 m²
Building Volume: standard freight container
Number of containers: 1
Construction costs: net construction and associated costs, theater building with "Musik der Jahrhunderte" approx. € 15 million, administration with restaurant approx. € 2 million
Building services engineering: air-conditioning and ventilation concept
Photographer: Dietmar Strauss / Bildermacher Architekturfotografie

RAUMLABORBERLIN
JAN LIESEGANG AND MATTHIAS RICK
Germany, www.raumlabor.net

EICHBAUMOPER // PAGE 130
Engineer's names: implementation planning: Jan Liesegang, Matthias Rick (both of raumlaborberlin), Jan Stauf (Gesellschaft der Holzfreunde), Christian Rick (Raantec), construction statics: Necati Elitas
Client: "Eichbaumoper" production company
Completion date: 2008
Total floor area: 50 m²
Building Volume: 160 m³
Number of containers: 4
Structural system: Container frame system with partial steel-structure reinforcements
Construction and materials: Interior insulation, wooden battens, insulating material, vapor barrier, drywall, simple planking
Building services technology: electrical, mobile counter/bar with water tank
Construction costs: € 70,000
Photographer: Matthias Rick

RAUMTAKTIK
FRIEDRICH VON BORRIES AND MATTHIAS BÖTTGER
Germany, www.raumtaktik.de

FANSHOP OF GLOBALIZATION // PAGE 89
Client: German Federal Center for Political Education
Completion date: 2006-2008
Total floor area: 15 m²
Number of containers: 1
Photographers: Nicolas Bourquin ((01), Willi Weber (02, 03)

RENNER HAINKE WIRTH ARCHITEKTEN
Germany, www.rhwarchitekten.de

CRUISE CENTER // PAGE 208
Engineer's name: supporting structure: Werner Sobek Ingenieure, Stuttgart, fire protection: HPP, Braunschweig, building services technology: HHLA, Hamburg
Client: Hafencity Hamburg GmbH
Completion date: 2004
Total floor area: 1,500 m²
Building Volume: 15,300 m³
Number of containers: 48
Structural system: supporting structure made of 20-foot and 40-foot freight containers; used containers stacked to form three layers; loaded with ballast to counteract wind and lever forces; extended using retrofitted steel frames in container dimensions; roof-supporting structure with wooden truss design, mounted on perimeter steel support.
Construction and materials: building with no climate shell; existing asphalt ground with bright red polyurethane coating; carpet in container, container with maritime blue-to-green tones on the outside; roof covering with steel sandwich covering and translucent twin-wall sheets.
Construction costs: € 1.2 million + € 0.6 million for site
Building services engineering: Indoors not heated
Photographer: Fotodesign Christoph Gebler

ROBERT DUKE ARCHITECT
WITH KIETH DOYLE, IAIN SINCLAIR, NICOLE MION AND EVANN SIEBENS
Canada, www.containr.com

CONTAINR CINEMA // PAGE 128
Creative Team: Robert Duke Architect, Kieth Doyle, Iain Sinclair
Artistic Director/Curator: Nicole Mion, Evann Siebens
Engineer's name: Bush, Bohlman & Partners Consulting Structural Engineers
Client: presented by Springboard in partnership with the Vancouver 2009 Cultural Olympiad
Completion date: 2009
Number of containers: 2

ROSS STEVENS
New Zealand

CONTAINER HOUSE (KILLER HOUSE) // PAGE 118
Engineer's name: Rob Bryant, TSE Group
Client: Ross Stevens
Completion date: 2006
Total floor area: 90 m² insulated, 45 m² covered verandah, 60 m² garage
Number of containers: 3 containers + 3 shipping platforms
Structural system: 3 vertically stacked containers elevated 3 m off ground level on re-used industrial crane sections

Appendix

Construction and materials: re-used steel, aluminium and polycarbonate
Construction costs: NZ$ 200,000
Photographer: Petra Alsbach-Stevens

SCULP(IT) ARCHITECTEN
Belgium, www.sculp.it

LIVE/WORK SPACE // PAGE 124
Engineer's name: Stubeco bvba
Client: Peerlings-Mertens
Completion date: 2007
Total floor area: 60 m²
Building Volume: 180 m³
Structural system: steelstructure
Construction and materials: wooden beams, wooden floors (parket), brick walls (existing), concrete floor
Construction costs: € 125,000
Photographer: Luc Roymans

SEAN GODSELL
Australia, www.seangodsell.com

FUTURE SHACK // PAGE 102
Completion date: 2001
Total floor area: 15 m²
Structural system: steel
Photographer: Earl Carter

SERDA ARCHITECTS
Austria, www.serda.at

SPACEMAN SPIFF // PAGE 84
Completion date: 1997
Photographer: Architect Alexander Eduard Serda

SHE-ARCHITEKTEN
ULRICH HAHNEFELD, MARCO PAWLIK, STEPHEN PERRY, STEPHAN SCHRICK AND TORSTEN STERN
Germany, www.she-architekten.com

ARCHITEKTURBOX // PAGE 160
Engineer's name: WTM Engineers GmbH
Client: Initiative Hamburger Architektur Sommer e.V.
Completion date: 2006
Total floor area: 85 m²
Building Volume: 250 m³
Number of containers: 5
Structural system: system structure
Construction and materials: material collage
Photographer: Oliver Heissner

SHIGERU BAN ARCHITECTS
Japan, www.shigerubanarchitects.com

NOMADIC MUSEUM // PAGE 200
Architect: Shigeru Ban Architects
Engineer's name: Buro Happold
Completion date: 2005
Total floor area: 3,020 m²
Structural system: steel container (wall)
Construction and materials: PVC membrane (roof, wall) plywood (wall), paper tube, water seal coated (columns) gravel, wood plank deck (floor)
Photographer: Michael Moran Photography, inc.

SHIGERU BAN ARCHITECTS + KACI INTERNATIONAL INC.
Japan, www.shigerubanarchitects.com

PAPERTAINER MUSEUM // PAGE 204
Architect: Shigeru Ban Architects + KACI International Inc.
Engineer's name: Minoru Tezuka, Myung-Hwan Lee (SAM-HYUN TG)
Client: Design House Inc.
Sponsor: Seoul Metropolitan City, SOSFO, SAMSUNG Electronics
Completion date: 2006
Total floor area: 3,820 m²
Structural system: foundation-steel beam frame 10 in container box + 40 in container box for roof paper tube column and paper tube roof truss
Photographer: Design House Inc.

SPILLMANN ECHSLE ARCHITEKTEN
Switzerland, www.spillmannechsle.ch

FREITAG FLAGSHIP STORE ZURICH // PAGE 122
Engineer's name: henauer gugler zürich
Client: Freitag lab AG
Completion date: 2006
Total floor area: 120 m²
Number of containers: 17
Construction costs: € 400,000
Photographer: Roland Tännler

TEMPOHOUSING / JMW ARCHITEKTEN
the Netherlands, www.tempohousing.com

KEETWONEN // PAGE 172
Engineer's name: Tempohousing
Client: Housing association De Key in Amsterdam
Completion date: 2006
Total floor area: 31,000 m²
Building Volume: 80,600 m³
Number of containers: 1,030
Structural system: self containing, no super structure necessary; steel shell (containers) have a special Tempohousing design (front and rear frame reinforced)
Construction and materials: stacked containers, steel-frame balconies and walkways and staircases, foundation on prefab concrete piles; additional steelframe roof trusses with insulated panels
Construction costs: € 20.8 million ex VAT
Building services engineering: per 150 container homes, 1 container with heating boilers (natural gas fuelled) that feed all the individual containers. Temperature per container to be adjusted. Each container individual mechanical ventilation with automatic variable speeds. Each container a 4 group fuse box. Hot water per container with electric heated tank of 50 liters. Cooking on electrical stove. All units have their own bathroom (shower, toilet, sink) and their own kitchen.
Photographer: Tempohousing

TEMPOHOUSING/KERSSEN LIJBERS
the Netherlands, www.tempohousing.com

SKAEVE HUSE // PAGE 69
Engineer's name: contracted engineers for electrical, mechanical, structural and building physics
Client: AWV Housing association and De Key housing association in Amsterdam
Completion date: 2007

Total floor area: 180 m²
Building Volume: 468 m³
Number of containers: 6
Structural system: self containing, no super structure; containers have a special design (front and rear frame)
Construction and materials: just containers on the ground floor, founded on stelcon plates; additional stone concrete façade
Building services engineering: temperature per container to be adjusted. Each container has mechanical ventilation with automatic variable speeds. Each container has a 4 group fuse box, hot water per container with electric heated tank of 50 litres, cooking on electrical stove, all units have their own bathroom (shower, toilet, sink) and their own kitchen.
Construction costs: € 108,000 (container building only)

WOLFGANG LATZEL ARCHITEKTEN
WOLFGANG LATZEL, NORBERT ZSCHORNACK
Germany, www.latzelarchitekten.de

FLYPORT // PAGE 212
Engineer's name: Bearing structure: Herbert Fink, Berlin; Building equipment and appliances: Büro Kleemann, Berlin
Completion date: Planing since 2004
Total floor area: 4,200 m² (prototype)
Building Volume: 25,000 m³ (prototype)
Number of containers: 410 (prototype)
Structural system: Steel frame construction consisting of moduls
Construction and materials: Standardized passenger terminal; three standards have been developed: high, medium and basic standard, depending on the demands of the user or operator.
Building services engineering: Standard equipment in heating, air conditioning, sanitation, electric, additional airport specific systems
Construction costs: starting from 2,000 €/m² (incl. planing, add. transportation)

YASUTAKA YOSHIMURA
WITH MANABU MIZUNO
Japan, www.ysmr.com

MIRROR ERROR // PAGE 90
Architect: Yasutaka Yoshimura
Art Director: Manabu Mizuno
Client: House Styling
Completion date: 2005
Total floor area: 30 m²
Building Volume: 77 m³
Number of containers: 1 40-foot container
Photographer: Daici Ano

Appendix

EDITORS

Prof. Han Slawik

Diploma in Architecture at the Technical University of Braunschweig/D 1973. Architect in several offices. Scientific assistant at the Technical University of Dortmund. 1984 Professor at the Hochschule/University of Applied Sciences Coburg/D, since 1994 Professor of design and construction at the Leibniz University Hanover/D. Architecture studio's in Amsterdam/Almere/NL and Hanover/D. Experimental buildings and projects. Numerous international publications, prices/awards, lectures about "*Container*Architecture". Participation Architecture Biennale Venice 2004.

Dipl.-Ing. Julia Bergmann

Born 1971. Degree in architecture from the Technical University Braunschweig/D 1999. Work for different architecture offices. 2003 establishment of an office in Berlin/D. Reasonably priced single-family homes made of steel in cooperation with the office kleyer.koblitz.architekten. Various awards and recognitions. Since 2004 scientific assistant of Prof. Slawik, Institute of Design and Construction, Experimental Design and Construction Section at the Leibniz University Hanover/D.

Dipl.-Ing. Matthias Buchmeier

Born 1977. Degree in architecture from the Leibniz University Hanover/D 2005. Since 2005 freelance work as an architect for different architecture offices. Since 2005 scientific assistant of Prof. Slawik, Institute of Design and Construction, Experimental Design and Construction Section at the Leibniz University Hanover/D.

Dipl.-Ing. Sonja Tinney

Born 1974. Studied abroad in 2004/2005 at the Universitá IUAV di Venezia/Italy. Degree in architecture from the Leibniz University Hanover 2006. Freelance work for different architecture offices. Since 2006 collaboration in diverse container construction projects at the office of Prof. Han Slawik. Since 2008 scientific assistant of Prof. Slawik at the Institute of Design and Construction, Experimental Design and Construction Section at the Leibniz University Hanover/D.

Imprint

CONTAINER ATLAS
A PRACTICAL GUIDE TO CONTAINER ARCHITECTURE

Edited by Han Slawik, Julia Bergmann, Matthias Buchmeier,
and Sonja Tinney
Co-edited by Lukas Feireiss for Gestalten

Texts by Han Slawik, Julia Bergmann, Matthias Buchmeier,
and Sonja Tinney
Project descriptions by Shonquis Moreno for Gestalten
The History of the Shipping Container by Kitty Bolhöfer for Gestalten
Guest contribution *Structural Engineering and Freight Containers*
by Douwe de Jong

Cover motif: Platoon Kunsthalle Seoul by Platoon
and Graft Architects, photography by Platoon
Layout design by Birga Meyer for Gestalten
Layout by Natalie Reed for Gestalten
Typefaces: Bonesana by Matthieu Cortat,
T-Star Mono Round by Mika Mischler,
Generika Mono by Alexander Meyer
Foundry: www.gestaltenfonts.com

Translation by Patrick Sheehan
Proofreading by Lyndsey Cockwell
Project management by Julian Sorge for Gestalten
Production management by Janine Milstrey and
Vinzenz Geppert for Gestalten
Printed by Optimal Media GmbH, Röbel/Müritz
Made in Germany

Published by Gestalten, Berlin 2010
ISBN 978-3-89955-286-7
5th printing, 2015

© Die Gestalten Verlag GmbH & Co. KG, Berlin 2010
All rights reserved. No part of this publication may be reproduced or transmitted in any form or by any means, electronic or mechanical, including photocopy or any storage and retrieval system, without permission in writing from the publisher.

Respect copyrights, encourage creativity!

For more information, please visit www.gestalten.com.

Bibliographic information published by the Deutsche Nationalbibliothek. The Deutsche Nationalbibliothek lists this publication in the Deutsche Nationalbibliografie; detailed bibliographic data are available on the internet at http://dnb.d-nb.de.

None of the content in this book was published in exchange for payment by commercial parties or architects; Gestalten selected all included work based solely on its artistic merit.

This book was printed on paper certified by the FSC®.